FACETS OF EROS
PHENOMENOLOGICAL ESSAYS

FACETS OF EROS
PHENOMENOLOGICAL ESSAYS

Edited by

F. J. SMITH

and

ERLING ENG

MARTINUS NIJHOFF / THE HAGUE / 1972

ISBN 978-94-011-8384-0 *ISBN 978-94-011-9078-7 (eBook)*
DOI 10.1007/978-94-011-9078-7

Table of Contents

Introduction   VII-XIII

1. Enzo Paci (Milano), "A Phenomenology of Eros"   1

2. Wolfgang Blankenburg (Heidelberg), "The Cognitive Aspects of Love"   23

3. Manfred S. Frings (Chicago), "The 'Ordo Amoris' in Max Scheler"   40

4. Don Ihde (Stony Brook), "Sense and Sensuality"   61

5. Erling Eng (Lexington), "Psyche in Longing, Mourning, and Anger"   75

6. Henri van Lier (Brussels), "Signs and Symbol in the Sexual Act"   91

7. Maurice-Jean Lefebve (Brussels), "The Nude as Symbol"   101

8. F. J. Smith (Chicago), "Don Juan: Idealist and Sensualist"   116

Introduction

In an age which is supposedly experiencing a sexual revolution, a volume of thoughtful essays on eros is not only not out of place but perhaps is a positive contribution to the understanding of contemporary man. It was the conviction of the editors that the scientific view of sexuality, as promoted in such valuable studies as those conducted by Masters and Johnson, needed considerable supplement and perspective. The perspective is here furnished by writers from both Europe and America, authors from various fields, such as philosophy, psychology, and even musicology, all of whom are united, in that their approach to the problem of eros is phenomenologically oriented. At first it might well seem strange that musicology would have much to say about eros. It is true, musicology has been the "science" of music, at least in intent. Yet in a larger view of the discipline, philosophical and aesthetic problems are also important to it, and this particularly if we agree with Enzo Paci, that our very culture depends on eros. Surely musical culture, as pointed out by Kierkegaard, is the embodiment of what western civilization has known as sensuality; and Mozart's *Don Giovanni* is its incarnation. On the surface it is easier for us to grasp the work of the philosopher in this area; and, of course, one expects the psychologist to deal with sexuality more explicitly than anyone else.

Enzo Paci gives us the Husserlian base for any discussion of eros in terms of phenomenology. His insightful article opens up the horizons of eros for us, in that subjectivity and bodily intersubjectivity are thoroughly analyzed to and including the act of coupling as the supreme expression of intersubjectivity. This fleshly union is a felt need of man, but he feels it, phenomenologically speaking, only from the depths of absolute solipsism. For man as utterly alone feels that lone-

liness and dire isolation that literally *impels* him toward the embrace
of the other. And thus he is said to have a sexual "impulse." Yet this
impulse is not just physical; it is bodily in the sense of Merleau-Ponty.
I tend toward and am impelled out of myself toward the other, in
order to overcome the abyss of solipsism and discover a mutual world
of love. Eros is thus "intentional," and it is this teleologically significant intentionality which gives meaning to and constitutes meaningful intercourse. The *logos* is already incarnate in the sexual impulse, as
the *ingathering principle* of unity, which causes two people to come
together in eros. They literally gather one another together in the
embrace of loving union.

Paci stresses that sexuality is not the obvious matter we commonly
think it to be, but rather that, making use of the reductive method of
phenomenology, we have to discover the sense of this enigma, and
thus the sense of intersubjectivity itself. And there is a profound bond
between signification and the bodiliness of "corporeity." For "intersubjectivity" is not just an abstraction or something arrived at in in-
tellectualistic manner alone; it is the bodily bond between two lovers.
The tendency to postulate a bodiless *logos* is the error of a typically
abstract intellectualism. But it is precisely the strong bond between
meaning and body that characterizes genuine sexual life. In this life
the world, as it were, surges and wells up in my being, and to embroi-
der the thought of the author, it is this *surgissement* which is the best
description of eros. For eros is the welling up and the surging forth of
one human subject in the direction of the other, of one bodily being
toward an other, as the lovers fall into one another's embrace and in-
gather one another in the act of love in "living presence" to one
another. Here intersubjectivity is intercorporeity, the intertwining of
two bodily beings, the inter-course of two lives and loves. Of this fruitful union results not only a shared world but new life. And it is brought
about in the rhythmical, i.e., as it were, the musical, composition
being effected by the lovers.

At this point I may be allowed to refer to what Don Ihde in his
essay calls the "field state" of sensation and sensing, which is achieved
in the mutual embrace. In ordinary terms this "field state" is *pure
sensation,* and in the erotic embrace prior to the act of love the harmonic atmosphere of the love-symphony is set. The feeling of pure
sensation is experienced by the infant held against its mother's breast,
by the little child that runs to its parents for warmth and to be "held."
It is felt by any human being or animal that luxuriates in the scent

and feel of a meadow in the spring. Here, of course, the *telos* is not love making. But this same pure sensation is the necessary prelude also to the act of love, if it is to have atmosphere and significance. For in the preliminary "play" the scene is set for the contrapuntal and rhythmic action that ensues. The emerging themes and counter-themes can be convincing only if the harmonic atmosphere of pure sensation is present: the feeling of "melting," of "floating," of being "carried away." And in the subsequent act of love the rhythmic undulations are not merely aesthetic, in the sense that we speak of the *art* of love; phenomenologically, it is a question of achieving *pure rhythm*, as the shared *life-movement* of the other.

Max Scheler strikes this writer as the most human of the major figures in phenomenology. He has the warmth and emotion in his writing that both Husserl and Heidegger seem to lack. It is hardly surprising we would find that for Scheler the "order of love" is prior to the order of being and of thinking. M. S. Frings, the general editor of the collected works of Max Scheler, presents the colorful thinker's views on love and its fascinating relationship with resentment. For Scheler the act of love is an act in which we discover basic value, basic attraction and repulsion, a whole world of value. In such a world love and resentment seem to go together. In this dialectic, however, there is ample room for confusion and for aberrant love, as in the case of Faust and Don Juan. Everything seems to depend on the direction of the act of love, and one can speak of a "topology of the heart." In the world of love and hatred we seem to partake of the *echos* of a deeper experience of world. The expression, *Echos der Welterfahrung*, appeals to a musician, especially one who, as a phenomenologist, looks for the sound metaphors in the major writers. The *echos* of Scheler is a "sounding of world-experience," in the sense that we sound a river to plumb its depths. Sound seems crucial to phenomenology, but the use of visual metaphor in the tradition has masked out attempts to treat of sound systematically. Scheler's position, as presented by Frings, seems to speak to our decade as to no other, perhaps not even the one in which he wrote. The essay is so timely that it even deals with the use of narcotics as one of man's attempts to do something about the joylessness of his existence. Treatment of the problem of being, of anxiety, of resentment, identify Scheler as a writer close to Heidegger. But he is more than a thinker who mediates between Husserl and Heidegger. He stands in his own right as one who has formulated basic phenomenological problems in a wholly original way. As a man he is said to have been most impressive. Apparently he lived the *ordo amoris* and did not merely write about it.

As to Don Juan, the idealist turned sensualist, the unacknowledged child of the ascetic, the present writer has contributed a lengthy essay, which deals with the intriguing subject of the Don in a manner more literary than systematically phenomenological. It is this musician's view that Kierkegaard's essay on Mozart's *Don Giovanni* is perhaps the single most important essay on the matter, unless we want to substitute mere musical analysis for a thorough understanding of what lies behind the character of the Don. In treating of Don Juan it was also deemed necessary to deal with allied matters, such as being and anxiety, Wittgenstein and Heidegger (on "world"), and many other matters that seem to influence our understanding of eros. But the essay speaks for itself and little more need be added in an introduction such as this.

F. J. S.

II

These essays are "evidences of Eros" in the sense that, as elucidations of the interplay of knowing and loving, they also give evidence of a transcendental plane through which knowing is discovered as grounded in a prior gift of self, an initiating gift, in and through which presence is effected. Descendant of the historical doctrine of divine creation, in which this bestowal was understood as a divine self-sacrifice enacting creation, it is now the *Lebenswelt* of Husserl which calls on us to recognize the reality of the *sinngebende Akte* (Scheler) of an apriori Love, a return of the repressed "cosmogonic eros" of the early Greeks. Here the choice of eros emphasizes the bodily (hence *physis*), even if the bodily dimension is transcendentally grounded, i.e., re-cognized through consciousness as evidence of an earlier self-bestowal already constituted, as Paci shows in his Husserl essay. In the Husserlian enterprise, analysis of what is revealed within consciousness as constituted leads back, through mediate intentionalities, to the possibility of an image of transcendental unity in which self and world already coalesce, often symbolically represented as a divine sexual couple or as a divine Androgyne. The "evidences" of the following studies are literally "outlooks" of that divinely poietic "inlook" (cf. G. M. Hopkins' "inscape") through which creation was conceived as having originally arisen. Now (as in that strange drawing ascribed to Bosch where features of the landscape are endowed with eyes) what is contemplated – as, for example, the nude, the coital pair, or the obscene – "looks out" through its particular form in a reversed perspective as if betraying its constitution as the self-disclosure of a transcendental eros. Within the occasion of this awareness, self and other are revealed to each other as concomitant disclosures of the correspondence itself.

If phenomenology is a manifestation of the concern to know things as they are, without injury to the eventual knowledge of who "I" am and "we" are, then it follows that it must sooner or later take up the relations of knowing and loving. It is in terms of just this connection, but, to begin with, in terms of the way in which loving participates in knowing, that Wolfgang Blankenburg undertakes to examine a central focus of the oeuvre of Ludwig Binswanger. In theological-historical terms (which the author eschews) it is apparent that if creation be understood as the accomplishment of divine ideation through which its love is embodied, then the human contemplation of things and persons in their otherness could restore them to terms of the originating theophany, i.e., to the *eide* as transcendental, and therewith, retroflectively, to the surprising discovery of a moment of transcendental love as grounding the possibility of our own being, as human and as individual. In one form of modern parlance this would be to realize the full sense of eros as "the unconscious." As already suggested, it can also be seen as the *telos* of the Husserlian *epoché* in the transcendental reduction. Thus it is possible to interpret the meaning of that "poetic phenomenology" broached by Michel Foucault (after Vico) – and interestingly enough in the context of Blankenburg's discussion – in his lengthy introduction to the French translation of Binswanger's paper on the dream. Such a "poetic" phenomenology can now be seen to derive its authority and force from the presence of the *eide* grasped not merely as the arch-forms of creation, but as symbolic evidences of the moment of eros which has guided our re-embodiment of them with immediacy and presence. In accomplishing this circle we extend originary creation, even while it is the pre-sense of creation which grounds, wittingly or unwittingly, the constituted partial autonomy of "myself" and "ourselves," as otherwise expressed in the "ritualization" of Julian Huxley and Konrad Lorenz, and especially in Erik Erickson's development of "ritualization" as the complement of "epigenesis."

The larger circle of creation, sometimes identified with the cosmos, has been worked out within the microcosm of the sexual partners in the lapidary paper of Henri van Lier. Here the *zarte Empirie* of Goethe is abundantly exemplified. The ambit of the couple, with its manifold possibilities of return and exchange, is shown as affording the possibility of the most intricate and inclusive knowledge of "actuality" or "identity as process" (Erickson). Through an attentively articulated *Wesenschau* the unity of consciousness and body emerges as *bios* transparent to the other anthropocosmic cycles. Pursued as it is into the physio-logical symbolisms of coitus itself, we are here offered a hermeneusis of physis as humanly sexual.

At the heart of the tendency to confine oneself to the most "private," i.e., "de-prived" circle, to become the "prisoner of sex," there is the suffering of a lived *aporia*, whose sense is illumined as if by a lightning flash in the myth of Diana and Acteon, the latter torn to bits by the goddess' hounds for having spied on her while bathing. Eager to grasp and to possess, he violated the necessity of realizing an initial gift of himself, his dismemberment an enacted symbol of the inseparability of self-bestowal from entrance into the secrets of creation and the divine. The theme of nudity becomes, in the reflections of Maurice-Jean Lefebve, a riddle about the relations of the private or *idios kosmos* with the public of *koinos kosmos*. Octavio Paz has provided us with the possible solution to this riddle given through "touch":[1]

> My hands
> Open the curtains of your being
> Clothe you in a further nudity
> Uncover the bodies of your body
> My hands
> Invent another body for your body

"Psyche in Longing, Mourning and Anger" probes the opacity of more encompassing horizons of understanding toward that search for meaning inherent in loving, as we note in the tendency for "knowing" to embrace all the distance between the intellectual and the sexual. Knowing has now shrunk to the level of mere curiosity. The tie of knowing and loving has foundered in an unstable antinomy of curiosity and abandon, mysteriously opposed even in their evocation of one another. That curiosity wishes to see without being seen, that abandon yields without giving; only the tantalizing repetition and circularity of the ordeal are a reminder of possibilities beyond the fission of longing, mourning and anger. In this there has been a virtual reversal of the

---

[1] *Hudson Review*, XXI (3), 1968 (Autumn), p. 460.

situation in which the cosmic, identified with the transcendental, has priority. The universe itself has come to have the characteristics of an enlarged human, a "Makranthropos," while the body has come to contain the cosmos. Thus the cosmic has become the anti-cosmic, now found within man, whose self-engendered suffering takes on overtones of an elusive transcendance.

A "phenomenology of eros," which this collection constellates, recovers for our time what was earlier known as divine love. Now, in the Husserlian perspective, eros is disclosed as a transcendental condition of all knowing, insofar as it is through eros that "transcendental ego" and *Lebenswelt* alike are to be comprehended in their apriori correlation. For anything at all to be, a self must have already expended or "emptied" itself, endowing the world with presence, just as in the earlier cosmic view the things and creatures of creation were seen as evidences of a self-distribution of divine being. It seems correct to understand this as love in a transcendental sense, i.e., as eros. The yielding of one's presently known life, as the bodily transmission of more inclusive constituting and constituted selves, is an initiation into the realization of the transcendant as continuing embodiment. This is the *telos* or goal of sacrifice, which is perhaps a feature of all rites as they contribute to the sustaining of transcendentalia. In this manner, every act from the contemplation of beauty to sexual congress may be grasped as sacrifice, as in the Brihadaranyaka Upanishad (VI.ii.13; iv.), a commemoration and enactment of our participation in creation.

E. E.
Winter 1972
Chicago/Lexington

ENZO PACI

A Phenomenology of Eros

I. OPENING TOWARD THE HORIZON OF EROS

The fundamental experience of phenomenology is *the reduction to subjectivity*. I must reduce myself to my self, to that by which I am "alive," to that which is my *Erlebnis*, to the experience I have *in the first person (in prima persona)*. Subjectivity has nothing to do with abstraction or with "categories." When I say "subject," I employ an anonymous term. In reality the subject, as unique, is myself, here and now, I who am writing. For reasons of method (but in the end not only of method) I have to begin with myself, and I can commence only with myself. It is not a question of giving an exposé of someone's thought or of constructing a philosophy; rather, I have to subject myself to an exercise that in a certain sense is *against nature*. It is "natural" for me to think of myself and to regard myself as a "thing" in the world, as an object in the world. My natural situation hides me from my self, "obscures" me. I live obscured as an object among objects in the world and am opaque to myself. If I have recourse to the sciences, they also treat me like an object, for they have forgotten their own origins, their own subjective "operations." If the sciences are "in crisis," it is particularly because of their much touted objectivity, their naturalism, because of that "mundane" naturalism, that characterizes for me my very being as lost amidst the mundane, my *Weltverlorenheit*. "Naturalism" and "objectivism" are "negative" terms used by Husserl in his tireless struggle against forgetfulness, the obscuring, the loss of me to myself, against vanity, the mundane.

The conquest of myself, the uncovering of that which I really am but which had been covered over, comes under the rubric of "reduc-

Translator's Note. Certain difficulties in translation, when felt to be significant enough, will be commented on in brief footnotes. Otherwise, the translator will not apologize for or comment on minor changes in punctuation or in style, necessary to render the "sense" of the author more properly in English.   FJS

tion." At this very moment in which I am writing, even if *in fact*, i.e., in that world in which I actually live (have always lived and will continue to do so), I am writing for someone, for a reader, *in truth* at the commencement of the phenomenological exercise I am alone, and I reduce myself to the unicity of my ego, to absolute solipsism. Yet this solipsism need not be one in name only; rather, it is an exercise, an *ascesis*. The difficulty of this ascetical exercise, the equivocations and paradoxes it engenders, are an integral part of phenomenology. The incomprehensions, to which the phenomenological exercise is prone, are infinite and would require a literature vaster than that of phenomenology itself.

If the disciples of Göttingen transmuted the sense of transcendental phenomenology into a Platonism of essences, not less serious has been the incomprehension of a reduction to subjectivity, of a transcendental subjectivity. I alone am subject, uniquely me. The world, i.e., all that which I feel and perceive, everything which I experience, of which I have *Erlebnis*, and whatever "is valid for me" in the modes in which it is "valid for me": it all is finally resolved in the "modality" of my subjective life. Whatever is not valid according to the modality of my life, my hopes, my understanding, my reflection, my knowledge; whatever is not *directly* experienced in its first-person evidence, which characterizes only that experience which is directly *present* and evident for me; all of it is not "lived" by me, and I am obliged to put it in parentheses. In addition to this I must exercise suspension of judgment, the *epoche*. Thus I enter within the "singularity" of philosophical solitude. From now on only that is valid which reveals itself to me in the first person, that which is a *phenomenon* for me, that which has significance for me.* The purpose of the *epoche* is, of course, a process of uncovering. The world is always there. *Its existence does not constitute a problem.* The problem the *epoche* must resolve is another: what is the *significance*, the *purpose* of the world for me, before all else and in original manner, and then for all such subjects? Insofar as it is lived in undeniable evidence, in all possible modalities as well as in all possible degrees of evidence, the world becomes significative, i.e., a bearer of meaning.* That which was hidden, closed, forgotten, offers me its meaning, becomes a giving-of-sense, *Sinngebung*.

---

* The Italian *significato* (*-ivo*) admits of various English meanings, from the word, meaning, itself through signification to significance. In each case the translator has chosen the English word which seemed to him to bring out the phenomenological sense intended or implied.

Reduction to subjectivity can go under the label of a reduction to the Cartesian *cogito*. Now, notwithstanding the continual admonitions of Husserl, the *cogito* is always a concomitant. The *cogito* is not the basis of thought; it is not the *res cogitans*, as for example my "soul", which is not extended, in contradistinction to the extended "body." *Cogito* is what I experience directly. It is that which gives of itself in me as sensation, as the sensations of all the organs of my body, and it is that which gives itself to me, of which I have experiential evidence, as a modality of my feeling, of the entire feeling of that organ of all sense organs, my body. In localizing itself and moving itself, in its feeling of flow, in self-experimentation, as with a thing that can be directly moved and guided in space and time; taking up, continuing and renewing past movements, in forgetfulness and in awareness, in obscuring and in making present again, in waiting and in anticipation, in imagining and in seeing beforehand – in all this I live in a complex network of experiences, in the flowing current of lived experiences, in an *Erlebnisstrom*. I live and I experience that which is truly mine, what is evident and pregnant with meaning for me. This is the way such persons as a Proust or a Joyce live. The world, not merely what "is said" to be the world, not the world in the testimony of others, but *my own proper world* is that which gives itself to me in this manner, whether in the *stream of consciousness* of Joyce or in the *Erlebnisstrom* of Husserl. In me myself (in a long process that is my own personal history, the unity of my story, of my own "novel") there is differentiated the imaginary from the real, the possible from the impossible, what is true from what is false. And it is in me, finally, in the varied ways in which I inhabit the world and live in it, before I ever reduce it to a concept or to a *category*. It is in me, in my precategorial life (*Lebenswelt*), that life which gives birth to all the categories of philosophy and of the sciences (including the categories of cause and of substance). When Joyce brings us back to the modalities in which we "live" the world, as it is valid for us and has significance for us, he gives us back a world already present and invites us to recoup within ourselves this original world in the freshness of its spontaneity. The world for us, the world in which we always are but which we ever forget and obscure, is the whole world of human beings.* Whoever knows this wholeness understands Husserl's insistence on the solipsistic exercise. For in the end I am not alone and isolated. There is al-

* *tutto il mondo*, the whole world, means colloquially „everybody." Here the philosophical sense of the whole world of humans seems to be stressed.

ways that which in principle is not only not "lived" by me and is not the subject of my personal experience but is in fact outside of me. In principle, whatever cannot be experienced by me has no sense for me. The solipsistic reduction looks to totality. Precisely because I reduce myself to my *uniqueness*, and *only if I do this*, everything is in me: the cosmos, the *Weltall*, is present in my uniqueness, in my individuality, in my singularity. Everything is in me, not in a spiritualistic or metaphysical sense, but in the most experiential sense possible to imagine. For I am the *Weltall* in my feeling, in my perception, by living in *this unique body* that I can call my own, in my own proper body. *A priori*, and here we deal with a material apriori, my body is in relation with all things. And this is the foundation principle, never comprehended in its concreteness, of *relationalism*, that principle of the *apriori of correlation*, which Husserl in a famous note in his *Krisis* says he discovered toward the end of 1898, the clarification of which was the purpose of his entire life.

That unique body which is my own, my body proper, my *Leib*, is the unique body lived directly by me, and through it I can say that *I am my body* in the vitality of all its organs, of all its *kinaestheses*, of all its achievements (*Leistungen*). In this my body and in its life there dwells the cosmos. It is there obscurely, potentially, but alive and functioning. In me are all things, the things of nature, the plants, as well as other human beings. When I uncover and unveil the things of the world and other living beings, I discover the *sense*, the signification which they have for me. I discover the world not as hidden but as revealed, as become *phenomenon, logos*. It is in this profound sense that, taking up Husserl's thought (*Ideen* I, 126), one can say that "logical" signification is expression. The *logos* is the *expressed world*, the world which has meaning for me, the subject, and finally for every subject as part of a society of subjects. What reveals itself in me and becomes expression, *phenomenon*, is that which gives itself to me in its true sense, that which acquires a significance. A thing that had no meaning becomes meaningful in transforming itself into a phenomenon. And that special thing, which the body of someone else is, in becoming phenomenal becomes *the signification of the body of the other*. As the existence of the world is not a problem, in like manner the existence of others is no real problem. I am always in the world, whether I like it or not. I always live with others, together with and counter to others. But what is *the significance of others* for me? And, reciprocally, what is my significance for others inasmuch as they are subjects? What I experience as the unique subject alive, as a subject which has overcome itself in self-reduction, is that which is valid for me, that toward which I move, to which I direct and open myself. In a word, it is that which is *intended* by me. That toward which my life, as subjective, intentional and transcendental, is directed is *intentional signification*. Thus the other is *one who has received signification*. Now how do I experience this signification? What are the modalities of my experience through which the other becomes significative through me? It is in replying to this question that I must reduce myself to the level of absolute solipsism, to that methodical solipsism, as it were, an artificial method that puts me in the condition of having potentially in myself all that is, yes, of having potentially in myself also the other. I must somehow produce the other from myself and thus gradually discover through myself his meaning and my own meaning through him.

The processes just indicated, if they are possible, are so, following Husserl's *Fifth Cartesian Meditation*, only if in that totality which is in me and which I am, I begin to discover what is *proper* to me; for it is precisely in this discovery that there can be any sense in my speaking of others and of anything outside myself. Now that which is proper to me is before all else my body, that live body, that *Leib*, which is me alone. How do I "live" the other in me, in that which is mine, solely my own? What are the modes in which I experience the other as alive in myself? How is there revealed in the total solipsism, that I am, that aspect of the totality which constitutes the other, the other in the very womb of what is proper to me? These questions, and the stupor such questions give rise to, if they are comprehended, experienced, and lived, open up to us the horizon of the problem of eros. Sartre and Merleau-Ponty, using Husserl as their point of departure, have given us the first analyses. We will bear them in mind, even as we return to Husserl in order to retrace the way and to follow it through. But it is indispensable that sexuality present itself not as something obvious but rather as an enigma, as a *Rätsel*. And it behooves us to be aware of the fact that man ignores, hides, and obscures *the meaning* of that depth experience which sexuality is, the experience responsible for birth and which is decisive in life and death, to a great extent for hatred and love, falseness and truth, for the very sense of one's history.

II. THE WAY THAT LEADS TO TRANSCENDENTAL PHENOMENOLOGY, BEGINNING WITH THE PHENOMENOLOGY OF EROS. STATIC ANALYSIS

I who live phenomenologically am a subject capable of intentionality. I am a *cogito*. Such intentionality begets signification, a *cogitatum*. If I am a *cogito* in my concreteness, in the reality of my being a "concrete monad," that which is signified is somewhat "unreal" and "imaginary." Intentionality once defined as "consciousness of something," this something which is intentionalized appears at first as unreal. The thing signified is not a thing but an idea, the image of an end, a *telos*. It is in following this indication that Sartre sees in consciousness the nihilation of being. If the other is an image which I make of the other, the other is illusory. If at a certain moment it no longer seems illusory, it nevertheless remains for Sartre the movement of negation, as it were, the movement of double negation of sadistic and masochistic coupling. In the first case I tend to negate the other through my own affirmation, in the second I negate myself in order to affirm the other in my own negation. The reduction of the other to thing, to object, to fetish, allowing only to myself consciousness and subjectivity, is a movement certainly tied to the dialectic of master-slave, as analyzed by Hegel in *The Phenomenology of Spirit*. The connection between this problem and developments of the Hegelian Left, even though they may seem obvious, or rather precisely because they seem obvious, are more profound than this.

The problem of the other and the signification of the other becomes almost insoluble for whoever remains with the statement of the intentional relation as *ego cogito cogitatum*. In reality from the very start the problem of intentionality presents itself in Husserl in various aspects. But it remains that intentionality tends to reveal the significance of the other for me, and finally the signification of subjectivity, of *societas*. And it is also true – and this is already evident in the *Cartesian Meditations* – that the significance of the other presupposes the discovery of my own bodily life and the reciprocal sensing by me of the other and of the other in me. It is in fact that experience which Husserl calls empathy or *Einfühlung*. In this experience the other is not experienced as alive* by me, as an "unreal" signification, as an idea, as an image of his own or of my life, unless he is first sensed as present in me, and I feel myself present in the other. Subjectivity is thus bodily, and re-

---

* *vissuto,* "lived" by me, a phrase that is not idiomatic English though good enough Italian.

flection on subjectivity in quality of expression and significative word, does not take from subjectivity its corporeity, even as it looks to the signification of corporeity. This profound bond between corporeity and signification, between materiality and *logos*, is a link which can give us a first revelation of the enigma of sexuality.

If I conceive of a consciousness minus body, the enigma remains unsolved. If I think up a type of consciousness, which does not experience within itself the world as alive, which is not a way of sensing the world or is not the world in the modalities of sensing, I rehearse the tired dualism of soul and body, of *res cogitans* and *res extensa*. The world of which we speak here must be the world that I experience, the world lived by me in the first person, not the "mundane" world on which I have practiced my epoche. It is in this world that whatever surrounds me, including material nature, is experienced as alive by me and is present in me, whether as limit, as *passivity*, or as conditioning causality. The world is my environment (*Umwelt*), that world of which I am a node of the modality of feeling as a sentient ego. A relation of activity and of passivity incarnates me in my *Umwelt*. That which conditions me is lived as conditioning, and it reveals itself to me thus in its significance, without which the conditioning would thereby cease. That which I suffer I continue suffering, even though I know "how" I suffer. And precisely this "as" can in its turn indicate to me active operations that "can" transform passivity into actuality.

Reflexion does not annihilate corporeity, because it functions within our bodily condition. Signification, even though it be a sign, a word, does not negate the thing itself, because it is itself the sense, the *logos* of the self-same thing. The signification of the other, precisely because it is the significance of the other, does not annihilate my experience of the body of the other. The *logos* does not detach itself from embodiment, because it is not "another thing" but rather is the actual meaning of embodiment. The tendency to destroy the flesh in the *logos* is a typical error of intellectualism; the opposite experience is the destruction of the *logos* in the body, thus objectification, fetishism (and the prejudice that sex is the more sexual when deprived of meaning). This dialectic is perhaps more original in the interplay between sadism and masochism. The other is the other body; yet this does not mean that it cannot be significative. Consciousness is always life in the world, and signification is really one of the elements which characterizes the sexual life. If we agree with Merleau-Ponty, it is only because the preceding clarifications permit us to correct his point of view: "Erotic perception is not a *cogitatio* that aims at a *cogitatum*; beyond body it envisages another body and it is effected in the world and not in consciousness." The point to be retained is that for Husserl consciousness is always embodied, and always in a concrete monad. The transcendental ego is not a mythological construct: it is me myself in flesh and bone, here and now, in that totality which, whether hidden or manifest, is alive in me. There is no *cogitatio*, which is not in the world and which does not experience in itself the world in all its modalities, as a material thing, as an organic body, as an animated organic body, as the life of the spirit, i.e., as social and cultural collective (the "objective" spirit of Hegel). The world is given to me *already* and *for always*, i.e., it is always a *vorgegebene Welt*. The *epoche* permits me to put it in parentheses, insofar as it is not experienced or lived by me, and permits me to sense it and to behold it singing forth in me, to live it directly in such a way that it can manifest itself and become *phenomenon*.

In subjectivity to which reduction leads me back, the world becomes a togetherness of modalities of my sensing and turns out to be in this very way my own proper world, the world lived by me, a world in which I find others experienced by me as alive. That which is lived by me is only what is experienced as alive by me in corporeity, in consciousness and in reflection; and it reveals itself, in this sense becoming immanent reflection on bodiliness as lived, thus imparting to me significance but giving it to me *precisely because the reflection is incarnate and individual*. The other is experienced as alive through that which is, but also through what can be, what in the future can become actual with me in a common *telos*, in a communal life meaning. Insofar as the other tends to steer his own life in reference to mine, and insofar as we tend toward a common goal, this purpose, as yet not realized, is an image, an idea, a wish, a hope or an agony, a transposition of lived experience into teleological image: it is the anticipation of the future, of a significant future. In the present this future negates and even recoups the past; our life continues renewing itself and remains in its one significative unity, in the possible significance of world, in its constant adaptation and self correction. If I am "transcendental" it is because I transcend myself, i.e., go out beyond myself with the other and with others in the continual perduring and renewing of the meaning of world, as constituted in myself and in the other, in me and in others. The other and I myself, we are necessarily a presence and a transcending of presence, a living corporeity and our own purpose, a perceptivity and an "image," a reality and a dream, a concrete life and a "logical" experience significative of such a life. The body experiences the other body as alive but likewise the *cogitatum*, the signification of the other body. It experiences the other as alive in itself and projects itself through sensible analogy in the other, while it feels the other projected in itself. I discover this original bond only in phenomenological reflection, in the reduction to solipsistic subjectivity. In reflection I discover now that which was already in me in the first place and continues to remain alive in me.

Like a philosopher who *has* the age compatible with philosophizing, it is I who now reflect. If I were but two years old, I could not engage in phenomenological reflection. The ideal of phenomenology is maturity, and maturity is the time for phenomenological reflection, a time that for many may never come. Maturity is also the ideal of normalcy; it is the highest point that the expression of *logos* achieves in its individual incarnation. Civil society is the vehicle of communication, the *reprise*, the renewal of the most significant and highest forms of communication, and this means of the "logical expressions" that are taken up again and renewed through new meanings in accord with the infinite teleological idea of reason, in accord with the idea of a human society comprised of thoughtful and self-constituting subjects, subjects for whom no subject is an object or slave. This teleologically significant intentionality is alive in *eros*, which gives and constitutes meaningful couplings (*Paarungen*), from which are born all the forms of civilization and culture. Significative and teleological intentionality is alive and already there covertly in the sexual impulse, which means that in the sexual impulse the logos is already incarnate. If I reduce myself to my solipsistic and bodily subjectivity, I find already there in me an intersubjective impulse, one that reveals intersubjectivity in my solitude and which is constituted in me as a need to get beyond myself into the other.

In my philosophical reflection, in the situation of static phenomenology, there arises the problem of the constitution of the other, the problem Husserl confronts in the *Fifth Cartesian Meditation*. I commence with the discovery in me of that which is proper to me, which is proper to the life of my ego; and I distinguish this from that which is *extraneous* to me. Since in solipsistic reduction I am the outsider, other than myself, I am like another me, an *alter ego*. My body is "proper." This body that I experience as alive and which can be touched by my bodily person, which completes the action of touching, as when with one hand I touch the other, this my body is both touching and touched, feeling and felt. In the primitive solipsistic ego, when I discover myself in that body which is uniquely mine alone, there is already some sort of distinction. There is the I, always the same I, as feeling and felt. Feeling myself I discover myself felt, and as I am felt I feel myself as a sentient being. I discover myself in an original sensible self-consciousness, in the evidence of the basic sense of touch. I "intention" myself. I am a subject which "senses" itself, and in feeling itself reveals of itself. Here the revelation is immanent to the sensation, in which self-consciousness is immersed. My body was asleep. But now it awakens and recognizes itself as bodily. It is now aware that it was also a body when it slept, even though it realizes this only now. Here there exists a deep knowledge, undifferentiated, fleshly, which reveals the immanence of the *logos* in the flesh, of "science" in sensibility and in bodily sensuousness. The original sexual discovery is that of my own body's feeling itself. This means that the sexual impulse strives from the start for meaning, for the reduction of the world to signification and toward that end which the world can have for me; it strives to reduce the world to subjectivity. I then discover that the same transcendental epoche is inborn in the bodily impulse of sensation and of sexuality.

In the same moment in which I experience my subjective impulse there reveals itself to me the possibility of becoming an object to myself, thus the possibility of fetishism and of objectification. My body can become an object of pleasure, the object of my subjective pleasure (e.g. in masturbation). Here in some manner I make an object of myself and become an outsider to myself, making myself a slave to myself while I am yet master. I alienate myself from my self. I feel myself as an object. I know and regard myself as an object (I who am rather a subject). Through this possibility of objectification and of fetishism *eros* announces itself as "sin." The object I become to myself is lost in the world, becomes a thing, and I want it to become a thing, a thing other than me, a life *extraneous* to me, a life that is in fact mine and solely mine. Scientific objectification is no different, when science forgets its own constitutive operations and its origins in the *Lebenswelt*. The *logos* becomes abstract, separates itself from precategorial life in which it became flesh; it forgets its own expressive, constitutive, "logical" operations. (Hence the profound connection between science and "sinfulness," a connection that is valid both for an unfounded science and for an alienated sexual life.)

In reality what I look for in my body, in the bodily circularity of the sentient-sensed, is something other than me, the *alter ego*. In a solipsistic situation the world does not exist as a world valid for all subjects, for all men. I can build a world, "objective" in this sense, only in communion with others in mutually verifying self-correction. In a solipsistic situation the entire world is for me only as the *other ego*, that other ego which I search out in my own bodiliness. I feel in myself an initial detachment, even when the hand which touches the other hand is sentient-sensed as it handles the other sentient-sensed member. I seek out the other as *alter ego* in my bodily solitude. The other must become me and other than me. We can only point to the conclusion which in reality would result from a long philosophical analysis. The *alter ego* is feminine, if I am masculine; masculine if I am feminine. But I myself am both masculine and feminine in my self before being so in others.

What happens if I touch the hand of another person, of another live body, another *Leib*? It will be *extraneous* in contact with mine, which is my own. It presents itself, it seems, as sensed, and I do not sense it *in me* as sentient, even though by analogy I guess it to be sentient. I extend my hand for example to a hand I see without seeing the body to which it belongs, like the hand that Anne of Austria offers D'Artignan to be kissed through a half closed curtain. I can project on this hand the feminine qualities which I sought in myself, and I could do so even were it the "feminine" hand of a man. What I imagine concerning the other I simply constitute, just as I have been constituted in the imagination of others. The imagination takes the place of what I seek, of what is lacking to me. It is already a tendency toward meaning, though toward an imaginary and presumptive meaning. My femininity projects itself onto another who can receive it, who is receptive to my projection, and generally, if I am a man, it is a woman. It is the entire world as *alter ego*, as femininity, and the whole world as that *other ego* which I look for in myself. The true sense of my own body and of my individuality constitutes itself in *detachment*, a disengagement in which what was mine, my femininity, projects itself into the outside. Thus the encounter with the other is such inasmuch as it is at the same time real *detachment*. My fantasy, my projection, becomes in truth the other body, the strange body, the very one that I have sought in myself. Once again I must find this self which has detached itself from me, make it become mine. I must make it my own and become my own *possession*. I have to possess it in order to reconstitute myself in my integrity, because it is *a part of me that has detached itself from me* and has become extraneous. I am a stranger to myself.

Yet in the impulse that leads to possession there is repeated the possibility of alienation, of objectification, of fetishism. The other can no longer be a fantasy but has to become objective reality, as mine, as possessed by me. She must become an object; at the extreme limit she must lose her subjectivity and die in me (aggressiveness, sadism); she must become my slave. In fact, in this alienation I lose myself and the other. The other whom I sought for in myself I actually looked for as subjectivity and not as objectivity, as felt and yet as feeling. I desired to reintegrate myself in the subjectivity I had lost. I wanted it to be other but as another "me," in the other as "mine," feeling thus in the first person in the other's body. *Eros* tends toward the constituting of a coupling (*Paarung*), of a dual life that flows from my original solitude. It is through this that I understand the other and her meaning, only if I reduce myself to solipsistic subjectivity.

Sartre and Ingarden reprove Husserl for his failure properly to constitute the other, and they accuse Husserl of idealism. But the so-called idealism of Husserl is *idealism of signification and not metaphysical idealism.* For Husserl the problem of the existence of the world is not a philosophical problem. The world is always already given, and a subject that does not have within itself a world is not possible. The problem is that of knowing what that meaning is which this world, as already given, can have for us. *To try to show that the world exists means putting oneself beyond the terrain of phenomenology.* The philosophical problem is posed precisely because the world is already given and is always experienced as alive by me. I myself am already and always given to myself along with the world I experience. To transcend oneself, to place oneself in *transcendental phenomenology* means that I cannot by an essential necessity remain that ego which I am as already given. It means the constant renewal and reliving of the significance of the world, of myself and of others, within me. The *logos* immanent in me must express itself, must be made explicit. This becoming explicit constitutes a *praxis*, a theoretical praxis and a life praxis. It is the concretization of every operation lived directly for a *telos*, toward a harmonic and rational goal.

To transcend oneself is to be "intentional," and this implies tending toward meaning. Before its being constituted as objective rationality (in the scientific sense) the *logos* unfolds itself as imagination and fantasy. Insofar as it leads to meaning, it goes beyond the real. The imagination is the source of all meaning, even if for this very reason it is also one of the founts of error, of procrastination, of failure to see

ahead, of deluded hopes. The exigency of signification is so prepossess-
ive as an impulse, in which the *logos* is immanent, that it bypasses reality itself, transcending it. In other words, *reality constitutes itself within
the imagination as one of its limits*, as a satisfactory place to rest, as a
verified assumption, as a fulfilled intuition. But, once again, within
the womb of this dialectic, alienation nests along with self-destruction
of the subject and its world. In its origins the fantasy of the masturbat-
ing adolescent is not different from the fantasy of the poet in the
"*dolce stil nuovo.*" Imagining is one of the modalities by which is made
present that which is absent, is lacking, which is no longer or does not
yet exist. Now that which does not yet exist, to the extent that it is
also what never will nor can be reduced to being, is truth. Truth is
teleological and infinite, and on this account it is inaccessible, in a
profound sense intentional. And it is that which *gives significative life to
being*, precisely because it never is being. The difference between
being and truth is insurmountable, insofar as it is indeed intentional
difference, the *difference between being and logos*. It is precisely this difference in potential that allows of every energy and of the teleological
direction of every energy and impulse. (In Heidegger this is all obscured, because the difference, which Heidegger calls "ontological,"
is posited between being and the existent essence; "truth" is wanting
as well as that intentional signification that can be constituted only in
subjectivity.)

The tendency, the direction toward truth and meaning is already
vitally there in the erotic impulse. Plato knew this. One need only
read the *Symposium* again. One of the deeper secrets of eros lies in the
fact that from the beginning *eros is directed toward truth rather than toward
reality*. This reveals the function of imagination in *eros*. (In Plato it is in
the final analysis *eros-in-idea*, and as such it is creation, *poiesis*.) This
also explains the fact that the final direction of eros is its orientation
toward a *society* of subjects, toward the ideal of a rational society. Yet
it also explains the loss of reality, illusion, failure, madness, or in psychiatric parlance, the loss of the reality principle. Finally it explains
revenge and resentment against truth, which then becomes a vendetta,
and resentment against reality becomes objectification, war, self-
destruction. The downfall of a civilization is closely linked with the
loss of the intentionality of truth immanent in eros. Here also the
"crisis" is due to the failure of the intentionality of the *logos*, to the
fading away of philosophical intentionality. This lapse is fetishism and
the sadistic-masochistic circle. Eros and torture find themselves in

Enzo Paci

analogous companionship. For torture is a modality of my feeling toward the other, and on his part of his own feelings toward me. In this case, recognizing myself in the other and the other in me calls for violence, and in the final instance collective self-destruction.

The immanence of truth in eros appeals therefore to human responsibility and to an awakening of conscience (*Selbstbesinnung*) in the functioning and in the signification of eros. In every copulation and in its modalities there is immanent the significance of humanity itself and its *telos*.

Intentionality does not make possession factually possible; or it reveals it as illusory. Not only this, but intentionality renders identification with the other impossible; and this means that on the intentional plane I am not the other, that the other is truly different from me and has to be experienced as alive by me as an other ego, as an *other subject*. If in the presence of the outsider I discover myself as I really am, in the encounter with an other subject which I am not, which is an other individual and an other person, I discover my self as an individual and as a person. The purpose of eros, of *Paarung*, is neither possession nor identification but reciprocal individuation, mutual discovery and maturation in becoming a person. Sexuality from the outset tends to individuate, not to make identical; in *Paarung* it tends toward mutual individuation and the reciprocal conquest of one's own personality in a distinct yet common life, in a life not merely dual but social, ruled by no master-slave dialect.

The attempt to identify and possess is thus objectification and fetishism. The master-slave dialectic can cease, however, only in a society where all are subjects and no one is an object. This is the *telos* of the transcendental reduction, of the reduction to subjectivity. This path on which we have taken but a few steps is the way which leads from eros to *transcendental phenomenology* (by means of that disclosure which reveals its sense, which carries eros forward toward expression, to the phenomenon). It is an aspect of the path which leads from initial solipsism to intersubjectivity. In myself I discover and experience the other as alive, as fantasy, as idea, as reality, as concrete individual, as person. In *Paarung* I find myself not through identification or in the pretense that the other gives me what I cannot give myself, but in mutual maturation, individuation, personalization. In *Paarung* there is constituted in original manner *societas*. I come forth out of my solitude only as a person among persons. But at the same time in a more profound sense *I accept my solitude*, inasmuch as with others I accept my self and reconstitute myself ever afresh. I may not free myself from that solitude which is the foundation of responsibility for self, of originality, of the non-repeatability of my life (which because it is irreversible is not repeatable and exists but once and for all, in this sense eternally), because I cannot and must not fly from my self, lose my subjectivity or pretend to lose sight of the subjectivity of the other. Intercourse (*Paarung*) is the primal nucleus of concrete intersubjective life, if the sense of life for the two subjects is not possession of either one of the two, but the *telos* of the one and of the other, finally, of the inter-subjectivity implicit in them both.

III. GENETIC ANALYSIS. THE CONSTITUTION OF THE OTHER AS PROCREATION AND UNIVERSAL TELEOLOGY.

As a concrete person who engages in philosophizing, and in order to do starts from the basis of self-consciousness, using my actual presence as a point of departure, from live presence, from the *lebendige Gegenwart* of my subjectivity, in investigating myself I conduct an investigation concerning the entire world which is alive in me. I can commence only from my actual ego localized in, become flesh in my own body, in my *Leib*. I am thus an original centrum that is knowledgeable and self-sensing, a point of origin for space and time. Everything that is alive is posited in relation to me, in material nature, in animate creation, in the life of the spirit. This ego that I am, the center of consciousness of self, expands in space and time, finds in itself what is near and far, the past and the future. I am a luminous center surrounded by shadows that fade into space and time. Obscure is that world which slips away from me spatially; and this is the world's history in its forgotten past and in its future not yet arrived. I can say that obscurity is that of both world and history, just as I can say that obscurity is in me, in my consciousness. In this sense a direct phenomenological analysis could arrive at the conclusion that the unconscious tends to coincide with the external world and with history as lived and forgotten, rather than with history as possible. In the last analysis the distinction between internal and external, between the subjective unconscious and the penumbra of the world, could show itself to be impossible or at least not obvious. Phenomenology's point of departure is consciousness, and in a strict sense one cannot speak of the unconscious. We thus have the dilemma of an unconscious, of which we have consciousness. On this plane the relation between that ego which I now am, which is capable of philosophical reflection, and the ego that I have been, is paradoxical. To be that ego which I am now, I had to mature, to traverse a path, which since birth has carried me along to the point of my now being what I am. If I start with that ego which I am and from this my center I illumine the obscurity that surrounds me (the obscurity which is in me) I move in the terrain of a *static phenomenological analysis*. But to move on this terrain, before ever moving in it at all, I have been genetically and historically formed. The ego, Husserl states, constitutes itself in its concreteness in the unity of its history. The ego which exists here and now can analyze itself in its own genesis, even if it be phenomenologically true that it can do so only starting with the present and in the present. The past is what is truly past, and thus it is a present past. But I can say this only now in the present, even though it be true that the past has not merely preceded the present but has formed it and has carried it along to become that which it now is. The present in its self-consciousness, in the clarification of itself, thus demands a genetic phenomenological analysis. The operation by which the present subject, which I am, clarifies itself, in order to know and analyze itself, to be fulfilled in its authentic maturity and to actuate a life style, a meaningful existence, is, to simplify matters, that operation which can be called *making present*. In this type of genetic phenomenology my purpose is to uncover the past, to bring to light what was forgotten, to make origins present and actualized. Paradoxically the actual ego, the I which I now am, the original ego, insofar as it is only from this I that our study can originate, precisely as an origin, is only in the present, in actual life, in *lebendige Gegenwart*. But the original I is also the ego which had a genetic origin, which was procreated. Living presence is never total presence; for, to be so it would have to make present in itself the origins of an infinite past, origins that are primary in time sequence. The originating process of the past moves forward into infinity and requires an infinite time for reflection, for discovery, for making present, a time that needs the future, an infinite future. Becoming totally present, origins move ahead, as it were, and become purpose, *telos*. Now precisely in time as actual, in actual presence, there exists the whole past and the entire future. In a hidden manner and in potentiality it is the temporal beginning of the genesis and the fulfillment of the genesis in the fact of becoming present. As E. Fink has observed, phenomenology can present itself as a problem, basically traditional, of the genesis of the world. It is a genesis, we note, which has always come about in the past, that takes place in the present and will always be there in the future. In every stretch of time from out of the past there is experienced as alive that present, that *lebendige Gegenwart,* which is alive now and will be alive in the future. The world is continually originating itself in every one of its temporal manifestations. Ideally, I should be able to weld together in the actual present the origins of the past with the fulfillments of the future. Indeed in my actualized present, in the concreteness of my ego, the horizons of the past and of the future lose themselves in the non-presentable, in the impossibility of utter revelation and of total foresight. If *lebendige Gegenwart* is the presence of the entirety, this total presence, implicit and obscure, is never fully clarified in any of its temporal moments, in any of the concrete lives in which it is expressed. The continuity of *lebendige Gegenwart* is broken in the discontinuities of consciousness, in momentary lack of attention, in the need to put oneself within a limited perspective, in the fact that the world is experienced as alive in quality of total horizon, yet always experienced "in pieces." And this comes about through partial thematization of a background which projects itself ahead of itself, precisely because it is felt as totality, as an unattainable *telos,* as *transcendence.* Transcendence is that which I feel as my own but beyond myself. The world, inasmuch as it cannot be made completely present, appears as transcendental. If I could begin from the basis of the unconscious, I would have to say that the transcendence of the world is a projection of my unconscious, an exteriorization of it. If I could start with the "external world," I would have to say that my unconscious is a projection in me of the transcendence of the world, of that total world I do not actually know; my unconscious would thus be an interiorization of the world. These remarks, which substitute for a longer and more rigorous analysis, are valid in every way for my genetic history as well. The genesis that has brought me to the state of being what I am is a kind of interiorization of the history of the world. The ontogeny sums up a phylogeny (in a phenomenological sense rather than according to the categories of biology as a science or in accord with its demonstrations). But this summing up is possible only through the discontinuity of genetic evolution. Continuity is broken by a pause; every temporal whole has its own rhythm, its own style, its own individuality. It is the same history, but as in individual histories that have their own unrepeatable singularity, and thus their own individuation. In every individual the same history remakes itself in a new way. That the same history is being dealt with makes analogy, type, and genus possible. It is a fundamental principle of phenomenology that in every individual it is possible to understand its type, its essence, its *eidos*. Genesis proceeds by types and individuals. *The type repeats itself in the unrepeatable individual.* My history is solely mine, but it is at the same time the history of the human race, universal history, that genesis which always has been and always will be, the ever present birth of the world. Type realizes itself in repeating itself, in renewing itself, in *rebirth*. The interruptions in intersubjective *lebendige Gegenwart* permit repetition and renewal of the same history, the individualization of genesis in finite periods of time, in individual lives. The continuity of presence is broken in the repetition of universal history, in every individual history. A hypothetically total self-consciousness, an absolute presence, would not be typical, would have no history, would neither repeat nor renew itself. In this case universality would coincide with singularity. In such a situation the world and I, all things and I, others and I would coincide in absolute identity. I would no longer be myself and one could no longer talk of solipsism. The solipsism of phenomenology illumines me as to the presence of every thing in my own singularity. But precisely because of this presence, through its concrete spatial and temporal modalities, it individuates me and thereby it places me in relationship with individuality and typology, with renewal and repetition, with continuous universal flow and its eddying in individual histories. The interruptions guarantee the same continuity and recapitulation. Pauses of this sort could appear as lack of a required universal consciousness, interruptions of life, death itself. And surely an analysis in this final direction is possible. But it is an unsuppressible principle of phenomenology, that only consciousness explains the unconscious and that only life explains death. Without life, without the presence of life, death is impossible. In its deeper sense life is originative, and in reflection it is the giver of meaning (*Sinngebung*). The life-logos thus understood never becomes totally obvious; flesh and logos, beginning and end, are never reunited in themselves in total identity. Because of this it is *intentional* through the impossibility of the coincidence of being and truth. Because of this the task of phenomenology is to make the infinite, which is forgotten, obscured, hidden, apparent and phenomenal.

In more concrete form, as a philosopher who inquires, I never succeed in making completely manifest my genesis, that genesis which comes about in me at this moment, that genesis which brought me about. I do not succeed in making present what has been forgotten from the very beginning. I do not succeed in reliving my birth at this very moment. The possibility of an absolute knowledge is brought up short before these breaks in continuity, in the face of that pause which constituted my birth and individuated me. This means I do not succeed in remembering my own birth as present. My babyhood days are immersed in forgetfulness, and a more profound oblivion closes off forever my intrauterine existence. As a philosopher who now reflects on the genesis of his own life, who tries to make present once again his own genesis, in going back into the past I find myself before a mystery. Directly, in the first person, I shall never know how I was born. This is the enigma of my own origin, which projects itself onto the enigma of the creation of the world. The idea of the creation of the world on the part of some deity is tied to this situation. It is a situation in which the problem of the creation of the world is confounded with that of the constitution of the other and with the problem of "how babies are born". In the *epoche* I have put all this in parentheses, insofar as it has not been experienced by me and I only know of it indirectly from others. On this account, from the phenomenological perspective the philosopher in his full maturity and in the august wisdom possible to him becomes a little child again, repeats his own genesis, his own history, the proper constituting of the world. But, just as the infant does not know how it was born, he finds himself confronted by something he cannot make present, i.e., his own birth, the transcendence of the world. I discover myself therefore as a child of the world, as *Weltkind*, as child of that world which transcends me, as a child of transcendence. In reality I am tardy in constituting also the so-called objective world in my dealings with others. The natural world of the sciences, for example, is the result of concordant intersubjective operations, always in need of the correction of a new concordance. The transcendence of which I am born is a thing of mine which is not mine, the figure of the *alter ego*. I shall be late in knowing who the mother is within whom I was formed. I will know that I was conceived within her through the insemination of my father; but before knowing anything, I have been alive, so to say, in the presentiment of this knowledge. And when I even find out "how babies are born," I will have another kind of knowledge, that can function and act, even if it does not become conscious and self-conscious. This knowledge is nonknowledge. It is the knowledge of the impossibility of rescuing from oblivion the moment of procreation, the "inwardness" of procreation, as Husserl put it. My body knows more than I can ever know of my body. It remembers what I can never recall or make present. Its life, as sexual impulse, is the impetus which makes it capable of recalling and repeating the "inwardness of procreation". It wishes to return to this inward region, yet realizes it cannot. The nostalgia for the maternal womb, for regression, is in conflict with renewal and repetition. For the Freud of after 1920 (after *Beyond the Pleasure Principle*) all of this presents itself as a struggle between the principle of pleasure and that of death. It is the same conflict which appears in Husserl as the struggle between subjectivity and objectivity or fetishism, and in Sartre as the conflict between sadism and masochism.

The sexual impulse in its normal functioning realizes itself in copulation and fulfills itself in reproduction, in other words in the encounter between immanence and transcendence in the birth of a child. But the moment of re-creation, the inward process of procreation remains obscure for me as a subject in the first person. It cannot be made present. More precisely, the internal process of pro-creation of which I am born presents itself solely in a new creative act, that in its turn will be capable of being made present not in memory but once again in a new creative act. From renewed presence another being is born of the "human race," a being which in that race can realize various essential types, first of all the female or male type (and this would require a considerably more profound analysis). That which is general and typical is what is characteristic of various planes of analogy (of various typical regions). History repeats itself and renews itself in the internal process of procreation, in heredity and in emergence.

It is a fundamental fact that the question of "how babies are born" is resolved only by making new babies to be born. There is here an essential upset that characterizes normal sexual life. The knowledge sought by us, which my very body seeks, is not actuated by the past but by the present and the future. The body knows what it knows and has always known only in that renewal which is for it a novelty, never before experienced, of the sexual act of which it is born. That of which the body has a presentiment is the completion of itself, which lies behind it in the copulation of one's parents, but which for it is fulfillment (*Erfüllung*) of its own presentiment, of its own *Meinung*, of the need to attain self-knowledge, to realize itself in its own completeness. Presentiment is expectation; expectation is prefiguration and desire, sexual impulse. The encounter of our parents and their mutual compenetration, the realization (thus also *Erfüllung*) of their coupling, of their *Paarung*, of their reciprocal feeling in interbodily *Einfühlung* (intersubectivity is in this case intercorporeity), all this, not presentifiable from the past, is upset in the future, in expectation and in the desire of a procreative future.

That which I seek in the evidence of the sexual act is at the beginning my own birth, the solution of the insoluble mystery of my genesis and that of the world. It is not the birth of a child but my own birth that never happened for me in the first person. In the sexual act I know myself as the birth of the world in me and I find once again the other in me and myself in the other. For me as a man the death instinct as instinct to the maternal womb becomes the penetration of the woman who takes the place of my mother and becomes a mother, mother of an other. This is the concrete *constitution of the other* which is based on my concrete subjectivity. In such a way the further encounter with the other becomes separation all over again. It is this detachment, already in sexual rhythm, that leads to orgasm, and which immediately after orgasm and satisfaction finds solitude once again. The impulse is not continual and infinite, and the pause that follows upon orgasm is a prefiguration of the more profound fatigue of life, of the weariness of death. Yet in the meantime within the unfathomable and non-presentable "internal process of procreation," in the enigma of the birth of the world, of the creation of the world and of man, birth has taken place, and a more profound separation has come about, the beginning of the life of an *other me.* Husserl describes all of this quite well, "The fulfillment of the impulse as penetration within the other animated life is not a reciprocal feeling oneself in the other as an uninterrupted experience of the life of the other, so that in the world it leads to that event which is in the world the act of procreation. The intersubjective act of procreation brings about natural processes in the other life. . . In the satisfying of the impulse seen in its immediacy, there exists nothing of the child that will be born. . ." (cf. E III 5, now printed in the appendix of my book, *Tempo e verità nella fenomenologia di Husserl,* Latera, Bari, 1961).

The act is accomplished and the woman's womb is made pregnant, and this is done in the woman in her placing herself in the situation of fecundity, maintaining and maturing it. The result is the *birth of the other,* of whom there was nothing whatsoever in copulation itself. This other am I myself, as one already born, as cast into the world. I will become myself by renewing the sexual act that caused my birth and by positing this same problem all over again through my own offspring.

Two provisional conclusions to bring this analysis momentarily to a close. First, genesis, which links me to my parents, binds me to the whole human race, to universal genesis. And it links me in anticipation to those who will some day come about. Others are in me and I in others. In the unpublished manuscript already cited Husserl speaks of "intersubjective impulse" and of "all the monads being with and in one another in the unity of a universal happening." Secondly, copulation and sexual satisfaction do not realize the erotic impulse, which insofar as it contains the logos within it, tends toward truth and can never come to rest in one being or in being as such. The purpose, the *telos*, is satisfying oneself with a fullness that is also separation, tends beyond sexual satiety toward the *meaning of truth* of intersubjective and social life, and it is expressed in works of culture and civilization. It is for this reason that the meaning of eros is transformed into the meaning of human history, or as Husserl states, into *universal teleology*, which is the same as saying that it leads to the final meaning of *universal phenomenology*.

transl. F. J. S.

Source: *Nuovi Argomenti*, nos. 51-52, 1961, pp. 52-76, "Per una fenomenologia dell'eros."

W. Blankenburg

The Cognitive Aspects of Love *

When affectivity, i.e. feelings and moods, are referred to in connection with the acquisition of knowledge, it is customary for the way in which they disturb cognition or even distort truth to be placed in the foreground of discussion. Knowledge is considered as mediation of what is "objective," in all the complex meaning of this word. By contrast, the world of feeling appears as the realm of subjectivity. What can it do but impair understanding? Thus it is usual to exclude feelings and moods as far as possible where knowledge is concerned, especially scientific knowledge.

Of course such an exclusion in absolute purity is scarcely possible. After all, choice of the direction of inquiry as well as of the preferred cognitive themes are also determined by affective tendencies, through a preference for this or that. A cognitively stimulating effect is readily granted to many emotionally anchored impulses as well as to curiosity. But that is only to begin with. The prevailing view is that where real knowing begins affectivity is out. Love in particular is reputed to be both prejudicial for understanding and corrupting of truth. It is not meaningless that there is a saying "Love makes blind." Who can contest it? This is a commonplace attitude, but apparently easier to uphold than its converse: "Love makes one see." If a choice were to be made today as to which favored knowledge more, sympathy or antipathy, there is no doubt that antipathy would be favored. For critical capacity – at least in our day – counts as one of the highest qualifications of the man of knowledge. And what is more favorable to a critical attitude than antipathy? And what is more likely to suppress criticism than love?

Nevertheless there is a tradition extending back into antiquity which ascribes a supremely significant role to love in the development of knowledge. The most striking evidence for their relationship is to be

* To Erwin Straus on his 80th birthday.

found in the Bible: "And Adam knew his wife Eve and she conceived
. . ." Here knowing is made synonymous not only with love in general,
but immediately with sexual love. The tree of (sexual) temptation is at
the same time the tree of knowledge. For centuries this has provoked
men's thought. There is a no less important than brief essay by Franz
von Baader: "On the analogy of the cognitive and procreative impulses" which, though more in the form of aperçus, elucidates these
connections. From here it might be possible to span the distance between pornography and spirituality in Henry Miller's work.

An intimate tie between knowledge and love is to be found in the
most various spiritual traditions of the past. Here the ancient mysteries
come to mind. Perhaps it is just this tie which provides an important
key for understanding their "esoteric" character. Basic for them is the
conviction that before valid knowledge is possible, knowing itself and
the knower as its bearer must first change. As the servant of such
change, realizing new forms of understanding, love has, from time
immemorial, been considered to have a special function. The obvious
goal is a transformation of the *ability* to love (at the same time it is
given a new direction) and not merely an access to love understood as
certain preferences or sympathies. Correspondingly in our century
Rudolf Steiner has demanded: "It must become possible for man to
make the ability to love into a cognitive strength" (1924, 50). The
widening of the range of possibilities of knowledge through the transformation of emotional abilities into gnostic ones as the fundamental
concern of spiritual traditions is a thesis that needs to be carefully do-
cumented historically and tested to see where it is applicable. If it
should prove correct, then the question would be whether this does
not reveal an affinity with psychotherapeutic training, insofar as it
aims at metamorphizing and refining empathy into an organ of analytic cognition. In this perspective psychoanalysis would appear as
secular successor to the esoteric traditions of antiquity. But we will not
pursue this further here.

In the history of Western philosophy love as an organ of knowledge
was (after Plato) most powerfully introduced by Augustine. Joining
Neoplatonic ideas with Christianity, he took the position that "Non
intratur in veritatem, nisi per caritatem." While Christianity first of
all simply divided the wisdom of the world as foolishness in the sight of
God from the Christian message, in an antithetical manner, thereby
creating the gulf between knowledge and faith which continues to the
present day, it was Augustine who made possible the question: "Why

should the search of knowledge be forever excluded from a life in which love has been raised to the highest commandment?" From Augustine a direct line passes through medieval Platonism to Pascal: "Il faut les aimer pour les connaître et (literally after Augustine) on n'entre dans la verité que par la charité."

In spite of this great tradition it must be observed that virtually everything written in earlier centuries on the relationship of knowing and loving remains confined to particular, and mostly aphoristically formulated, theses. Only from the basis afforded by the phenomenological research of Edmund Husserl has the relation between knowledge and love become scientifically accessible. To be sure, the purely scientistic concept of truth had already been shaken by Nietzsche. Knowledge as form of world-mastery: this postulate led to seeing "interest" as an essential condition of cognition (Habermas). A systematic treatment of the order of priority between conceptual (*vorstellende*) and interested (*interessenehmende*) acts occurs for the first time in Max Scheler. In his paper "Love and Knowledge" (1916) he made an express tie with the Augustinian tradition. He took a firm position for the primacy of acts involving interest, particularly the "primacy of love," over against knowledge (X, 370). In the "apriorism of loving and hating" he saw the "foundation of every other apriorism," defending the thesis that "In it, and not in any 'primacy,' whether of 'theoretical' or 'practical reason,' do the spheres of theory and praxis find their ultimate phenomenological tie and oneness" (II, 85). Before man is "an *ens cogitans* or an *ens volens*" he is "an *ens amans*" (X, 356). "All knowing of an object and all willing of a project are commonly grounded in the love of the matter common to this object and project" (II, 553). According to Scheler, the "intentional nature of empathy (*Nachfühlen*) and sympathy, as well as its function of providing material, is cognitive in the same prelogical sense in which the perception of concrete states of affairs is 'cognitive.'" Cognitive here means virtually the same as *world-opening*. Consequently epistemological questioning does not start with the problem of the certitude of a particular knowing, but is located much earlier, namely in the problem of *openness for* something, or in the problem of the conditions that make it possible that something is there at all *for* man. If Pascal, as we have seen, appealing to "les saints," pointed out that there are objects which, contrary to the popular proverb, must first be loved in order to be known (*erkannt*), Scheler establishes this as valid in a transcendental sense for all objects of knowledge. In so doing, love is established as the secret *movens* of all interest, a basic interest as foundation however for every cognitive event.

This problem complex undergoes a sharpened formulation, at the same time taking another turn, through Heidegger. In "Being and Time" (1927) he inquires, in the course of explicating the problem of the meaning of "Being," into the understanding of Being (*Seinsverständnis*), and in this framework into the openedness of existence (*Erschlossenheit des Daseins*) in man. Scheler and the rest of the tradition mentioned above he extends, insofar as for him it is one's condition (*Befindlichkeit*), above all, responsivity (*Gestimmtheit*) which co-constitute the openedness of the world: "The mood has already disclosed, in every case, Being-in-the-world as a whole, and makes it first of all possible to direct oneself toward something" (1927, 138; 1962, 176). Accordingly, in inquiring back beyond intentionality, a fundamental role for the pre-intentional disclosure of the world is ascribed to affectivity (in the broadest sense of the term): "Indeed from the ontological viewpoint we must, as a general principle, leave the primary discovery of the world to 'sheer' mood" (1927, 138; 1962, 177). With responsivity (*Gestimmtheit*) is meant not so much particular isolated feelings and moods but rather *that* world has already been opened in feelings and moods of whatsoever sort: "The responsivity of one's condition constitutes the openness to the world as 'Being-there' " (*Die Gestimmtheit der Befindlichkeit konstituiert die Weltoffenheit des Daseins*) (1927, 136). Heidegger is interested in the power of moods, and of affective impulses generally, as opening, not so much the mode of being of what is concretely encountered within the world, as opening, the "world" as the "wherein" of everything encountered. In that respect he is a more thorough-going transcendental philosopher than Scheler. And yet Heidegger not only proceeds explicitly on the basis of the inquiry concerning πάθη in Bk. II of Aristotle's Rhetoric but just as much on that of the tradition defined by the names of Augustine, Pascal and Scheler. In conscious contrast to Descartes, Heidegger, in elevating to the norm not the *certitudo* but the broadest possible disclosure of self and world, not only extends this tradition but at the same time exposes its foundation. Adhering to the "farthest-reaching possibilities for the opening up of *Dasein*" (1927, 139) he distantiates himself from the conception of knowledge oriented to the ability to doubt, to *claritas* and *distinctio*, which has dominated the West, at the very latest, since Descartes. For him "breadth" takes precedence over "certitudo." Consequently Szilasi (1946, 12; 1969, 71) understands Heidegger's philosophical activity as a grandly conceived attempt to "win back" for human thought "the most originary breadth of transcendance."

"Breadth" as criterion of the openedness of the world, and openedness of the world as criterion of knowledge and its truth content, has become axiomatic for the phenomenological direction of inquiry. As sole criterion, that would allow deviation into dubious biases, above all those of irrationalism and the glorification of immediacy, in contrast to every form of mediacy (Adorno). Not however if this criterion is employed as a conscious supplement to that of the *"certitudo,"* i.e., the dependability and precision of knowledge. Both criteria together, in their polarity and mutual enhancement, are alone able to do justice to the nature of truth and the dialectical process of its development.

But only if the breadth and depth of knowledge is taken as seriously as their certification does the question arise: what other functions of human existence, outside those of the sensory organization (viewed as encapsulated) and the intelligence referred to it, participate – or are capable of participating – in the disclosure of the world? This question – in the philosophical tradition *that* of the transcendental power of imagination – is fundamental with respect to a possible cognitive function of love.

In "Being and Time" one will search in vain for a transcendental interpretation of love; there are treatments of dread, fear, curiosity, etc., but none of love. That is not by chance. Behind it lies not only the concern that in such a discussion one might fall into sentimentalities. It can be shown that there are more principal reasons against it. (1) Heidegger is interested in the preintentional relation to the world; however, the prevalent view is that love is among those feelings of particularly clear intentional directedness. (2) Even more important is this: that love is not taken up in Heidegger becomes explicable if one considers that in a certain way it allows the infinite to flare up within the finite. Heidegger's point of departure however is particularly accented by its being oriented to the finite: responsivity (*Gestimmtheit*) does not disclose world as such – prior to every differentiation between finiteness and infinity of *Dasein*. From the very beginning Heidegger rather accepts the prior status of finitude. The responsivity of one's condition "opens up *Dasein* in its projectedness" (*Geworfenheit*) (1927, 136), and not at once projected and projecting. The pathic moment of feelings and moods (the way in which *Dasein* is required to live its Being) is regarded as their essential moment. The result of this is that dread becomes a "basic conditionality" and "Care" a basic existential disclosing *Dasein*. How it does this is not the question here. What is alone important is that "Care" and "Dread" are not invoked as psychological states, but only as they are, ontologically viewed, manifestations of the finiteness of *Dasein*.

The question of the original openedness of world is accordingly in Heidegger, just because of the role in it played by affectivity (under the heading of "conditionality" *[Befindlichkeit]* the affective is already placed in the framework of a certain ontological interpretation), tied up extremely closely with his thesis of the finitude of *Dasein*. This thesis, to be sure, appears explicable from Heidegger's starting position, as axiom even obligatory. But if taken as more than a heuristic principle, it turns into a dogma. Heidegger himself was evidently clear about that. In "Kant and the Problem of Metaphysics" he places his finger on this most vulnerable spot in the initial conception of "Being and Time." Giving a hint, as it were, to his critics, he asks there: is the transcendance of *Dasein* and therewith the understanding of Being actually "the innermost finitude in man," and does the founding of metaphysics as such have this innermost tie to the finitude of *Dasein* (214); finally: "Is it possible to develop the finitude in *Dasein*, even as a problem, without a 'presupposed' infinity?"

Here is the point which is open to attack, and from which Ludwig Binswanger, in his comprehensive work "Basic Forms and Understanding of Human Existence" (4th ed., 1964), sought to broaden the foundation of "Being and Time." Like Heidegger he seeks the broadest possibilities of the opening up of *Dasein*, by no means limiting these to the traditional cognitive functions, but even including the so-called 'Affectivity." Objective knowledge, as it dominates the natural sciences, is for him as for Heidegger a derived mode of *Dasein's* more original modes of disclosure. He differs from Heidegger in not going along with him in starting exclusively from the finitude of *Dasein*. According to Binswanger, *Dasein* can "be understood 'in' its finitude only from the infinite" (1964, 226 ff.). The structure of "Care" elaborated by Heidegger ought not to lead to the mistake of granting finitude an ontological priority over the infinite. In Heidegger's view (1951, 214 ff.), the Being of what is is only intelligible as such – and in that the deepest finitude of transcendance is to be found – if and when *Dasein* in its very depths holds itself into nothingness"; and he regards Dread as that basic conditionality in which nothingness is confronted. If in Dread the inner finitude of *Dasein* is manifested, Binswanger asks in turn: What discloses the infinite in *Dasein*? The answer he gives is

Love. If "Dread" and "Care" are transcendentally interpretable, i.e. as organs for understanding Being, why not Love as well?[1] To the ontology of "Care" he counterposed an ontology of "Love." Of course he had no intention of replacing the former by the latter. Heidegger's ontological explications were too evident to him for that. Binswanger rather took the position that only from an insight into the polarity of "Care" and "Love"[2] were it possible for the full transcendant structure[3] of *Dasein* to become clear.

Binswanger's opus has been much criticized. He was reproached, just as were the efforts of Bollnow aimed in the same direction, with an anthropological-psychological misunderstanding of Heidegger (Kunz 1941, 1949). This was so convincing that Binswanger finally, in the Introduction to the 3rd and 4th edition (1964), charged himself with this misunderstanding. It had been pointed out that "Care" understood as an existential is the fundament of love no less than of care and dread (now understood as psychological states). Rightfully so.[4] Yet

---

[1] In that case it is assumed that love is in no wise exhausted in "being directed to something," but that it is characterized over and beyond that by a pre-intentional reference to Being as a whole, just as is true in Heidegger for "dread" in contrast to "fear." This is however just what Binswanger attempts to show, with the help of abundant quotations from world literature. He goes even farther. Just as for Heidegger it is "Dread" which grounds "Being-in-the-world," for Binswanger it is "Love" which grounds a "Being-over-and-beyond-the-world." That is possible of course only if Heidegger's claim is contested, to have demonstrated with "Being-in-the-world" an all-embracing structure, for which there would be no "beyond," in contrast to which an "over and beyond" would be in principle absurd, because such "Being-in-the-world" grounds every "over and beyond" to begin with, and thus necessarily encapsulates it.

[2] Binswanger was not the first to see this polarity. Already in 1916, long before "Being and Time" appeared, Scheler had written: "The extreme opposite of earthly love is. . . care born of vital anxiety" (X, 308). Of course "care" and "love" in Scheler are, even less than in Binswanger, raised to a problematic level comparable to that of "Being and Time."

Ernst Blum (1945) in a study of the relationships between psychoanalysis and knowledge of Dasein which is still worth reading today, has compared the polarity of love and care with that of the sexual and ego instincts in Freud, as well as with that of Eros and Thanatos To draw such a parallel can however only be fruitful if the relations of the various levels of observation to one another have been clarified.

[3] The interpretation of transcendance as Being-*in*-the-world had already appeared one-sided to Binswanger (cf. Note 1). Consequently, for the full structure of transcendance embracing both Care and Love, he devised the unwieldy formula: "Being-over-and-beyond-the-world-in-the-world." To the transcendance of Care as "climb beyond" (*Überstieg*) he counterposed the transcendance of Love as "flight beyond" (*Überschwung*). This baroque terminology is not altogether happy; it undoubtedly owes something to misreadings of the Heideggerian intentions. And yet what it implies remains thought-provoking even today.

[4] It is appropriate for an existential just as for the transcendentalia, to be comprehensive. Inherent in that are the dangers of making it absolute and committing violence. An impression is created that what has been understood in the perspective of a particular existential is incapable of being interpreted – perhaps even more deeply – in very different perspectives. Here is a point where transcendental thought is threatened with succumbing to a universal power claim, possibly because fundamental for it is already a hidden tendency to obtain mastery. Which does not mean that it must succumb to this temptation. Binswanger has performed the service of having uncovered the problematics of power and powerlessness

the possibility of an ontological interpretation of love is not thereby precluded. The criticism is justified only to the extent that Binswanger was factually unsuccessful in developing an ontology of Love which could equal Heidegger's ontology of Care and with it form a new whole. For that Binswanger's philosophical strength was insufficient. It was unable to match that of Heidegger, having moreover to be wrested from everyday psychiatric practice. Further it should be considered that an ontology of Love definitely requires greater philosophical strength than an ontology of Care, if it is to be as rigorous.[5] That does not mean in any way however that the task Binswanger set for himself was not and is not justified, or even necessary. To have formulated it clearly remains to his credit. For what is important is the problem, and not the solution he gave for it.

In any event Binswanger's work is, in the estimation of Theunissen (1965, 439 ff.), the most radical attempt to date to realize an ontology of the "Between." Scarcely anywhere else has the necessity for overcoming the gap which lies between a one-sided constitutive transcendental philosophy and a positivistic empiricism been so clearly seen. Binswanger has the edge on "dialogical" thinkers such as Buber, Marcel, Löwith, Rosenzweig, Rosenstock-Huessy and others, in that he not only sought to reconcile the transcendental point of departure with the dialogic of "I" and "Thou," but over and beyond to mitigate the deep rift in all these thinkers between nature and existence by recourse to Goethe's "thinking through objects" ("*gegenständliches Denken*"). How far and at what points Binswanger failed in this immense task will not be gone into here (cf. Theunissen, Sonnemann, *et al.*). Here the concern is not with an interpretation or critique of Binswanger's many-levelled work, stimulated from the most various directions; neither is it concerned with clarifying his position in the history of ideas, a position powerfully rooted moreover in Western humanism. Only a few of his thoughts with a direct bearing on our topic will be taken up and discussed with regard to their implications for the relation between love and knowledge.

latent in the thought of "Being and Time" (4th ed., 1964, 147 ff.). This danger cannot be avoided by a blanket renunciation of transcendental-phenomenological interpretation and ontology, i.e. through confining oneself to discursive-positivistic thought, but rather through a thoughtfulness not susceptible to becoming fascinated by *an* absolute, one aspect taken as absolute, a thoughtfulness which has developed the strength to sustain a universe of diversified aspects offering themselves as absolute at one and the same time as they remain reciprocally inclusive or exclusive of one another.

[5] The case is perhaps similar to paintings of scenes from the New Testament, in which the visage of Christ is often least successful artistically, just because it makes the greatest demands.

The role acknowledged by Binswanger for love would not be a transcendental one, if love were discoverable only in human *Dasein* and not at the same time demonstrable as the condition of all discovery. In accord with the theme of the present study we are not so much concerned with the significance of love for *Dasein* as such as for the *knowledge* of *Dasein*. Correspondingly we have less to do with the more extensive Part I of the work ("Basic forms of *Dasein*") than with Part II ("The nature of the knowledge of Dasein"). If in Part I the thematic is primacy of love (and therewith of duality or of the "Between" prior to individual existence), Part II establishes how far love must also have priority as method. Not only *Dasein*, but the *knowledge* of *Dasein* as well – in an extremely broad, non-scientistic sense also termed simply "psychological knowledge" by Binswanger – has "its own ground and basis in the loving compresence of I and Thou," and that means at the same time: its *criterion*. Knowledge is conceived – though not exclusively[6] – as a form of encounter,[7] and measured according to love as the most intense form of encounter. Binswanger begins his exposition of the knowledge of *Dasein* with an excursus on the dialectical identification of reason and love in the passage of cognition to recognition in Hegel. The basic position is contained in an opening chapter entitled "Overcoming the contradiction of love and care in the knowledge of *Dasein*." A second chapter clarifies the position further in terms of the history of ideas, Binswanger developing "knowledge of *Dasein*" as continuation of methodic directions developed by Goethe for the knowledge of nature, and by Dilthey for the sciences of mind and man.

What sort of knowledge is it at whose cradle love presides? Binswanger writes: "This knowledge is image-laden, beholding ideas or

---

[6] Knowledge of *Dasein* for Binswanger is by no means grounded exclusively in the "openedness of heart." He acknowledges fully and completely the necessity for a discursive, object-oriented inquiry. Knowledge of *Dasein* should rather encompass both kinds of knowledge in their incommensurability, without confusing them. It is "not at all the 'common denominator' of the truth of the heart and theoretical truth, but it is the wager (*Wagnis*)... that that incommensurability can be overcome" (555). This incommensurability is not discussed away. Rather it represents the basic problem of the knowledge of *Dasein*, its questionability (489). It cannot be said that Binswanger succeeded in his wager of "resolving" this incommensurability. But who has? That Binswanger even accepted this wager in "Basic Forms and Knowledge of Human Existence" lifts this amorphous, overladen opus way beyond many others which stand up as more accomplished and "successful." In any event there has as yet scarcely been another work which has attacked the problem of the relation between love and knowledge – as an epistemological-ontological problem – on such a broad base.

[7] Binswanger distinguishes knowledge through love from that which is discursive, which he interprets as "handling" (*Nehmen-bei-etwas*), for example, grasping something in terms of its objectality or its dominability rather than as itself.

essences, and yet at the same time delights in facts" (118). What
analysis of conditions is supposed to do for the empirical sciences,
ideation for eidetic phenomenology, "imagination" is to accomplish
for the knowledge of *Dasein*. For "love neither 'thinks' logically-dis-
cursively, nor metaphysically-substantially, but imaginatively-specu-
latively" (114). Hamann already wrote: "Senses and passions speak
and understand images alone." But from where do these images
originate? How does imaginative knowledge emerge from love? For
the moment it remains obscure.

The imagination as method for knowledge of *Dasein* is merely
sketched in outline by Binswanger, and more indeed with regard to
what it is supposed to accomplish than in regard to what it actually is
and how it comes to be. As so often with Binswanger it is, considering
the major task falling to it, more celebrated than exactly described and
methodically analyzed. A clear grasp of the concept of imagination –
also in comparison with what is usually understood by this term –
would be exceptionally important however, not least in view of the
significant function Binswanger accords it. For after all this function
is supposed to bring about a reconciliation of the delight in facts with
the strength of the intuition of essences – thus the values of empirical
and eidetic research – closing the gap between factum and *eidos*,[8] facts
and essence, the individual and the generic, which runs through all the
history of Western thought, and beyond that even, closing the still
deeper gap between naive realism and transcendental inquiry.

What does Binswanger understand by "imagination"? In his own
words (156): a transcendental reality-'disclosing' mode of *Dasein*;
more precisely: "the transcendental unity of phantasy ('imaginality'),
revelation of reality and fancy . . . The imagination is clearly demar-
cated from empiricism as well as from phenomenological ideation.
From the former it is distinguished by the fact that it (like phenomenological ideation) transforms individual intuition into intuition of
essence. From phenomenological ideation imagination is distinguished
by the fact that it does not artificially exclude the natural attitude, but
elevates it, on the ground of a fresh naiveness, directly into the transcendental consciousness. Thereby it is both less and more than
phenomenological knowledge: "Less, because it is not descriptive-

---

[8] The "persistence in being restricted to the single fact or to the mere idea means the
closing up of the self . . . and the closing up of the world . . . in the hardening of the heart"
(Binswanger, 4th ed., 1964, 557). The accent lies on the "distance".

cognitive[9] intuition of essence,. . . more, because it is not *merely* (one-sided) intuition of essence in pure transcendental consciousness, but (two-sided) interpenetration of essences (*Wesenseinbildung*) in the realization of transcendental consciousness (648 ff.). Here essence is to speak to essence. The act of love which sustains the imagination is diametrically opposed to that of the phenomenological "epoche."[10] It permits the "impartial observer" to become a "participant."[11] That of which love, according to Binswanger, makes knowing capable, is "imaginative realization of the intuition of essence"[12] (645) and beyond that "realization of transcendental consciousness"[13] (648). What is the meaning of "realization of transcendental consciousness"? While the transcendental for Kant is the result of a reflection, for Husserl the object of an originary intuition, Binswanger calls for the transcendental to be realized, i.e., for its actualization, and therewith for it to be torn away from the apriori as everlasting "already," to produce it freshly in each case from the Here and Now. The transcendental (comprising the conditions of the possibility of something being present for us) accordingly no longer remains that which has only subsequently been laid bare (out of the noemata), but is to be newly developed in each case from the actuality of consciousness. The words "transcendental" and "apriori" undergo in this context a considerable change of meaning. Everything traditionally understood by these concepts becomes mobile and takes on a new face. Binswanger himself did not realize this change of meaning to its full extent. Else he would not have been able later repeatedly to recur to the Husserlian point of

---

[9] What is here meant is: critical cognition, capable of assuming distance to itself.

[10] A. Schutz and, following him, M. Natanson, have also described an act opposed to the phenomenological epoche: the "epoche of the natural attitude," to which man is considered indebted for anchorage in the life-world. It is to be read off from the resistance against which the phenomenological epoche has to contend (Blankenburg, 1971, 67 ff.). A special phenomenological inquiry would be needed, in order to ferret out the relationships between this "epoche of the natural attitude" and "love" in the Binswangerian usage.

[11] Here we might recall V. v. Weizsacker: "To make inquiry of what is living, one must participate in life" (1947, V), but above all the principle of "participant observation" of H. S. Sullivan.

[12] At this point a comparison with the more sharply conceived concept of imagination as employed by Steiner could take us farther.

[13] The aim to reconcile or even to unite empirical realism and transcendental consciousness is not new. Schelling had already established this as his goal in his "positive" philosophy. It is not by chance that it presents itself as a "philosophy of mythology and revelation. "The pursuit of this goal obviously leads to the problem of a transcendental power of imagination and to the problem of an imaginative consciousness in a virtually inevitable fashion. That explains why thinkers who take this path are always coming into the vicinity of myth and the poetic. This is particularly clear in Heidegger (cf. his interpretations of Hölderlin, Trakl, Rilke, George *et al.*, above all however his own poetic experiments in "From the experience of thought").

departure, failing thereby to understand both his own intentions and those of Husserl.

The heart of Binswanger's criticism of Husserl is his criticism of every form of unilaterally constituting intentionality. He replaces it with the in principle infinite process of the reciprocal constitution of object and consciousness. In this he follows in the steps of Goethe (617 ff.): "Every new object, well beheld, develops a new organ in us,"[14] in which context "organs" are to be understood as (transcendental) conditions of the possibility of experiencing. What turns simple looking into a "good" look, according to Binswanger, is love. With this the place is shown where love, in this conception, performs its function: it is a capability – perhaps the most powerful one – which makes us accessible to what is to be encountered, and indeed in a sense which is inclusive of the transcendental subjectivity.

Independently of Binswanger one can formulate the thesis: The world opens man up in order to be not only lived, but also vitally experienced – loved – and finally known through him. In its development the human self intensifies its subjectivity, not in order to be resolved into it, but in order to develop manifold organs of apprehension, by means of which the world is enabled to represent itself in ever more variety and depth on continually fresh levels of "objectivity." Here spontaneity and receptivity, project and "projectedness" (*Geworfenheit*) no longer stand sharply opposed; interest is directed rather to what is prior to these alternatives, in the form of *Anverwandlung, Aneignung, Einbildung.**

Binswanger understands the *imaginatio* also as imaginal consciousness or as imaginal cognition, to be sure, but even more as "loving identification." What does that imply? The word *Ein-bildung* does not

---

[14] In: "Meaningful advance through a single genial word" (1823). Also cf.: "Man, where he makes a meaningful entry, exemplifies the lawful. – Yet he doesn't always care to rule; he often prefers to abandon himself to love . . . In turning thus to nature a supremely happy situation arises: the mutual resistance ends; nature permits her deepest secret to become mysteriously transparent . .," in: "Problem and Reply (Essays on the General Principles of Natural Science)".

* *Translator's note.* Out of respect for both German and English and at the author's suggestion, I have chosen to let the original German words stand. Three progressive phases of a dialectically understood transformation of the self are described. The crucial word here is *Einbildung*. Generally, it has much the sense of the older "fancy" or of present-day "phantasy," when loosely employed. Literally, as *Ein-bildung* it would mean alteration of the self through assimilation of an image, as it were, an "endo-identification": thus the initial imaging of the other is succeeded by the appropriation of this as one's own image, which in turn is followed by a metamorphosis of the self. Binswanger's *imaginatio* which covers the gamut of possibilities, might be characterized as a "lived" imagination.   *E.E.*

mean that knowing here is to be relinquished to luxuriating phantasies. It is rather to be understood literally as endo-identification, and indeed in a two-fold sense: (1) as the entry of *Dasein*, as unending, through identification into finite being (156-157); (2) and here is where the emphasis is placed in the present context – as endo-identification of what is encountered into human comprehension for the purpose of opening up new organs of experiencing. Before we determine whatever is encountered, following our usual cognitive praxis, on the basis of a pre-established fixed schema of categories, we should open ourselves so far to the other in a process of inner assimilation (*Anverwandlung*), that the one who meets us is able to inform our categorial readinesses or schemata of comprehension, in order to, as it were, itself determine how it wishes to be known, queried in regard to what, etc.

How does the encountered wish to be conceived, how approached? What projects suit it, and which do not? How must I allow myself to be laid claim to by whatever meets me, so that I am able to address it, not merely in terms of its possibility of being dominated, but in terms of that through which it becomes capable of demonstrating ever more clearly what it is or is able to become? With these questions that which we encounter invades the organization of our transcendental subjectivity to a certain extent, freeing it from its rigid immobility. The aprioric character of our cognitive faculty loses a bit of its unrelenting apathy and with fresh immediacy re-enters correspondence with the apriori of nature. Necessary for that is a particular plasticity of the constitutive structures of awareness.[15] What kind of ability is it which produces this plasticity? In Hegel (23) it is the dialectical movement in which pure certainty (reason 'dead set'on itself) "surrenders the fixity of its self-postulation" and the static ideas become "fluent"; in Binswanger it is love. A new form of openness to the world is imparted to knowing through love; legislative reason is to learn to listen.

It is in this sense that the earlier cited motto "Love makes one see" is to be understood. According to Steiner (1908, 53) love is even the "sole passion which in the search for truth is not to be discarded," but simply transformed. Even if other affects – like aggressive impulses and antipathies, which are able to stimulate the ability to criticize, and also, above all, wonder – may subserve knowing, there is yet hardly anything which opens man up more powerfully in a positive sense for the world than love. Scheler already counted it among the "bestowing"

[15] Goethe: "There is a delicate empiric, which makes itself most intimately identical with the object and thereby develops into actual theoria" (Proverbs in Prose no. 906).

("*gebenden*") acts. Love is that force which most powerfully shatters the bounds of the self, certainly in what is at first a "selfish," i.e. subjective way. That is the reason why the statement "Love makes blind" has, at first glance, a specious justification. In the first moment, when love (emotionality in general) begins to activate the process of knowledge, misrecognition of reality, blindness of judgment, ideologies of every sort threaten in the most devastating manner, and indeed necessarily so. Nothing can "err so deeply, so fatefully, and so recalcitrantly as the heart" (Guardini, 1962, 121). Evidently the possibilities of penetrating into higher spheres of truth are to be had only at the price of just as deep possibilities of being deceived. One need not have any illusions about that.

But love is not on that account blinding, because less is seen under its influence. Rather what is new – often merely possibilities – keeps coming into view (for example, another's secret virtues, which otherwise would have remained hidden). Yet what is new emerges so strongly that other features are concealed, and the view of the total reality is disturbed or distorted. A person requires a very high degree of ego-strength to open himself lovingly to things beyond what the already formed intellect is capable of, to be able to expose himself without danger to the above-mentioned invasion into the organization of his transcendental subjectivity, at the same time permitting himself a greater openness and emotionally sustained enhancement of his sensitivity, without loss of critical capacity. It is a long, long way to free a feeling like love so far from every trace of self-indulgence, that the capacity for love becomes a capacity for knowledge, so that from it the strengths of a fresh openness to the world may grow, setting free previously missing possibilities of understanding.

No doubt it is "cheaper," because safer and at the same time more comfortable, to simply avoid the dangers of subjectivity, and with it a relation of love to the world, by banishing anything affective from the cognitive sphere as far as is humanly possible. Yet that could in the end prove very costly for mankind. As admirable as such an *ascesis* might be, and however great the control over nature and man thereby achievable, it is the equivalent of a cognitive *self-castration*.[16] And to

---

[16] This expression once again accents the analogy between sexual love and knowledge. Binswanger gives quotations from Rousseau, Goethe, and Kierkegaard, according to which the knower assumes a feminine relation to the idea, and then rejoins: "It is not the idea that 'fertilizes' or 'befalls' me, but 'the actuality' in which *I am* but as a member. The idea is already the fruit of this conception, its pro-geny." On the other hand, knowing also retains an actively masculine character. The relation of knowing to reality would be consequently *hermaphroditic*.

the castration of cognition a corresponding castration of the reality in which we are cognitively active. Human understanding thereby forsakes the possibility of any increase in depth. Increase in depth means disclosure of "deeper" levels of reality. In order to enter into them with understanding it calls for, not the elimination of subjectivity – and therewith of love – but for its purification. Purification means simply gradually to render subjectivity transparent, gradually transforming it into organs of experience, and at the same time, understanding.

The method of elimination continues of course to preserve a certain prerogative. This right is merely qualified. In affectivity, and thus also in love, there always remains a considerable residue, which must be kept out of the cognitive sphere, because its subjectivity cannot be made transparent. "To render subjectivity transparent" means in the last analysis that which is usually termed "subjective," but which contains the subjective and the objective as commingled; what is referable to world and what is referable to self (*Weltbezogenheit und Selbstbezogenheit*) are sharply distinguished. The method of eliminating everything "subjective," and therewith love as well, from the life of knowledge – till now the only recognized path to objectivity – thus becomes in no wise superfluous, though still decisively relativized. The method of elimination remains solely for the unpurifiable residue of subjectivity at each moment, the portion which cannot be rendered transparent and which still continues to have its significance in the whole of human psychic reality.

Viewed from this aspect, the collective organization of human understanding appears involved in a continual dynamic process. What today possesses categorical character for objective and even for scientific understanding, could very well have been wrung bit by bit from the subjectivity of psychic experiencing in the course of a millenia-long historical process.[17] The sphere of human affect, here represented by love, would then be a vast reservoir for future world-disclosing powers;

---

[17] To that would correspond the conception of Pascal, as suggested in Fragment 479. According to it we know "les premiers principes" not through the understanding, but through the heart. To the knowledge of the heart in Pascal belong not merely the religious truths but, for example, the tri-dimensionality of space, the infinity of the number series, etc., in short, everything of an axiomatic or apriori character. The "heart" – read affectivity, temperament, love – is consequently acknowledged as an organ of cognition, without of course necessarily implying the idea of evolution earlier suggested. At this point we may refer to Amiel: "Aimer c'est virtuellement savoir. . .," where in the present connection everything depends on what is understood by "virtuellement." – On the problem of subjectivity/objectivity as well as that of the intentionality of the affective life cf. H. Schmitz (1968, 1969) and most recently – with regard to the constitution of the life-world – G. Brand (1971).

it would contain a *prospective potential* with regard to an amplification of human experiencing.

This evolutionary theoretical hypothesis – providing it is correct – would shed light on that incommensurability between the truth of the heart and the truth of an understanding set on its reliability and demonstrability (cf. note 6) – in short, a light would fall on the relation between wisdom[18] and *science*. Transcendental inquiry – insofar as it can do nothing else but "hearken for the Being of what has already been previously disclosed ontically" (Heidegger 1927, 139) – can scarcely advance any farther here. It may investigate the conditions of the possibility through which anything at all is "there" *for* us (opened-ness of *Dasein*). It may ascribe to love as the highest degree of being interested in something an important function with regard to under-standing. But with that it nevertheless opens up no fresh possibilities of disclosure. The question of the conditions of the possibility, of that which "always already" must be presupposed, remains oriented to the past. But one does not do justice to love with regard to its possible, still latent cognitive function, if the possibility of a hidden prospective potency is not reckoned with, presupposing an evolutionary process of self-disclosure of the world, from vital experiencing to knowing, a process from which we ourselves are not excluded, however slowly its steps of development may be completed.

The present study cannot claim to have represented the relation between love and knowledge in all the width and depth of its proble-matics. Not only that thematically the question of the significance of understanding for love had to be bracketed out. In addition, the re-verse question of the role played by love, or called upon to play, in the world of knowledge, was only set forth, perhaps just far enough to stimulate a fresh systematic thinking-through of the problem.

[18] It may be recalled that "philosophy" means "love" of "wisdom." But love does not follow wisdom; it constitutes wisdom, transforming science into it.

transl. E.E.

BIBLIOGRAPHY

Adorno, Th. W. *Jargon der Eigentlichkeit*. Frankfurt a.M.: Suhrkamp, 1964.
Amiel. *Fragments d'un journal intime*, Paris, ed. Bouvier I, 27.
Augustinus. *Opera omnia* (Migne, tome VIII), Contra Faustum cited by author lib. 32, cap. 18, cf. M. Heidegger).
Baader, Fr. v. *Sämtliche Werke*; hrsg. von Fr. Hoffmann. Bd. I, 37-46. Leipzig, 1851.

Binswanger, L. *Grundformen und Erkenntnis menschlichen Daseins.* München-Basel: E. Reinhardt 4. Aufl., 1964.

Blankenburg, W. *Der Verlust der natürlichen Selbstverständlichkeit.* Stuttgart: Enke, 1971.

Blum, E. "Daseinserkenntnis und Psychoanalyse." *Mschr. Psychiat. Neurol.,* *110,* 47-67 (1945).

Bollnow, O. Fr. *Das Wesen der Stimmungen.* 2 Aufl. Frankfurt a.M., 1943.

Brand, G. *Die Lebenswelt.* Berlin: de Gruyter, 1971.

Guardini, R. *Christliches Bewusstsein. Versuche über Pascal.* München: dtv 38, 1962.

Habermas, J. *Erkenntnis und Interesse.* Frankfurt a.M.: Suhrkamp, 1968.

Hamann, J. G. *Aesthetica in nuce.* Schriften Bd. II. Berlin: 1821-43.

Hegel, G. F. W. *Phänomenologie des Geistes.* Leipzig: Meiner, 1907.

Heidegger, M. *Sein und Zeit.* Halle: Niemeyer, 1927.

Heidegger, M. *Being and Time.* English transl. by J. Macquarrie and E. Robinson. London: SCM, 1962.

Heidegger, M. *Kant und das Problem der Metaphysik* (1929). 2. Aufl. Frankfurt a.M.: Klostermann, 1956.

Heidegger, M. *Aus der Erfahrung des Denkens.* Pfullingen: Neske.

Kunz, H. "Die anthropologische Betrachtungsweise in der Psychopathologie." *Z. ges. Neurol. Psychiat., 172,* 145-180 (1941).

Kunz, H. "Die Bedeutung der Daseinsanalytik Martin Heideggers für die Psychologie und die philosophische Anthropologie." In: *Martin Heideggers Einfluss auf die Wissenschaften.* Bern 1949, p. 37-57.

Natanson, M. "Philosophy and Psychiatry." In: *Psychiatry and Philosophy,* ed. by M. Natanson. Berlin-Heidelberg-New York: Springer, 1969.

Pascal, Bl. *Œuvres complètes.* Bibliothèque de la pléiade, nrf. Paris: Gallimard, 1954.

Scheler, M. *Gesammelte Werke.* Bern-München: Francke, 1950 ff.

Schmitz, H. *Subjektivität. Beitrage zur Phänomenologie und Logik.* Bonn: Bouvier & Co., 1968.

Schmitz, H. *System der Philosophie III/2. Der Gefühlsraum.* Bonn: Bouvier & Co., 1969.

Schutz, A. *Collected Papers I: The Problem of Social Reality.* 3rd ed. M. Nijhoff: Den Haag, 1970.

Sonnemann, U. *Negative Anthropologie.* Reinbek: Rowohlt, 1969.

Steiner, R. *Metamorphosen des Seelenlebens* (1910). GA 59. 4. Aufl. Dornach, 1958.

Steiner, R. *Initiationserkenntnis* (1924). GA 227. Dornach 2. Aufl. 1960.

Szilasi, W. *Macht und Ohnmacht des Geistes.* Bern: Francke, 1946.

Szilasi, W. *Phantasie und Erkenntnis.* Bern und München: Francke, 1969.

Theunissen, M. *Der Andere. Studien zur Realontologie der Gegenwart.* Berlin: de Gruyter, 1965.

Weizsacker, V. v. *Der Gestaltkreis.* Stuttgart: Thieme 1947.

M. S. Frings

The "Ordo Amoris" in Max Scheler

Its Relationship to his Value Ethics and to the Concept of Resentment

We will cite a passage from Scheler's essay, *Ordo Amoris*, at the beginning of our investigation, which will give us the general framework of our theme. There we read:

Whoever grasps the ordo amoris of a man, has hold of man himself. He possesses him as moral subject – what crystal form is to crystal itself. He sees into this man as far as one can see into one's fellow. Behind the empirical multifold and complexity he sees the ever simply flowing basic contours of man, and this, rather than knowledge or will, deserves to qualify as the core of man as a spiritual being. He possesses in his spiritual make-up the original source that secretly spawns all that issues from man. And even more, it is the primordial determinant of that which incessantly places itself around him – in space his *moral world*, in time his *fate*, i.e., to become the quintessence of the possible that can happen to him and *only to him*.[1]

Although in the three periods of Scheler's thought, especially in the years from 1922 to 1928, there are many traceable modifications of previous writings, nevertheless his anthropological philosophy was always based on his fundamental position, according to which man is not primarily a knowing or willing being but an *ens amans*. For Scheler love remained the "mother and awakener" of all knowledge and willing, and for this reason the emotional apriori of love and hatred is the final fundament of any other apriori, i.e., of any knowledge of being or any willing of contents.[2]

It is amazing that in the literature on this great thinker one finds only occasional reference to the "order of love" in its central meaning,

---

[1] *Schriften aus dem Nachlass*, p. 348. When not otherwise noted, citations are taken from the *Gesammelte Werke*, Franke Verlag (Bern/Munich). *Editor's note*: M. S. Frings is now general editor of the complete works of Max Scheler. The translator of this essay has also agreed to translate vol. X (*Nachlass*). F. J. S.

[2] *Der Formalismus in der Ethik und die Materiale Wertethik*, p. 85. (4th ed.)

though directly or indirectly it permeates Scheler's entire writings.[3] Quite specific references to the *ordo amoris* are to be found in his *Formalism*, further mention is made in *Essence and Forms of Sympathy*, in his treatises on *Resentment, Knowledge and Work, Models and Leaders*, as well as in *Love and Knowledge*. An external criterion for the significance of the *ordo amoris* in the thinking of Scheler is to be witnessed in the fact that he treated intensively of this theme at the same time as he dealt with Non-Formal Value Ethics and resentment. The treatise, *Ordo Amoris*, remained but a fragment, and originally Scheler had intended to publish it with the title, "On the 'Order of Love' and its Aberrations." Manuscripts on this theme were lost, particularly the principal theme of the treatise, "Aberrations of the Ordo Amoris, Arranged according to certain Types, with their Origins explained."[4] There is no doubt that Scheler's study on *resentment* also represents research into the aberrations of the *ordo amoris* and that both treatises rest on his Non-Formal Ethics. Not only the above mentioned external criterion speaks for this but also the entire context of the problem, as it exists between the *ordo amoris*, resentment and value ethics. This is quite obvious from the many facets of Scheler's writings. Scheler had planned to elaborate on this interdependence for some projected chief works, but it was all aborted by his untimely death. For one cannot emphasize enough that Scheler wanted his Value Ethics understood only as the basis for elementary points of departure.[5]

In order to expound the central meaning of the *ordo amoris* for Scheler's value ethics and resentment, the *ordo amoris* should be sketched in its essential features. Everywhere Scheler indicates that value perception precedes any knowledge of being, since without the act of participating, one cannot have any *part* in being.[6] Every conative comprehension of an object presupposes an emotive experience of value, and this is equally true of perception, memory, expectation, and of all thinking. It is valid for every intuition of ultimate phenomena, as for all observation. Thus a child first grasps the value modality, "pleasant," before he grasps the quality, "sweet."[7] In similar fashion the love of one person for another is already *there*, before he encounters him intellectually or as an object of willing. The conceiving of a value always has the element of the mysterious about it, for acts of knowing

---

[3] Cf. the works of M. Dupuy, W. Hartmann, J. Hessen, H. Lützeler, E. Rothacker, *et al.*
[4] Cf. *Schriften aus dem Nachlass*, I, p. 516.
[5] Cf. *Der Formalismus*, preface to the first edition.
[6] Cf. also *Philosophische Weltanschauung*, Dalp, vol. 301, p. 40.
[7] *Die Wissensformen und die Gesellschaft*, p. 109 f.

and willing that are based on it seem closer to us in reflection, although
they are always guided from out of this background by emotive "value-
ception". Such guidance can be peculiar to one individual, a group, a
society, a nation, peculiar even to an entire cultural unit. In all men, as
individuals and as social beings, value experience, like some pioneer,
precedes every thought and act of willing. The peculiarity of a deter-
minate kind of "value-ception" consists for Scheler in the current
predominant rules of preference between values (and their change).
The experiential structure of "value-ception" (*Wertnehmung*) is a fact,
that stands behind a morality or a people in history (e.g., that of a
race); that is, the structure of feeling values in the emotional sphere of
man, and thus the preference or rejection of certain values, conditioned
by pure attraction or repulsion (love and hate), effect what with his
own peculiar expression Scheler calls "ethos," as constitutive for any
morality or type of ontological knowledge. Thus he writes:

> The most radical form of the growth and renewal of ethos is that uncovering
> and discovery of "higher values" (to the given) which takes place in and by
> virtue of the movement of *love*, and it is in the first place within the bounds
> of the highest of the value modalities. . .[8]

Morality and ethics (not philosophical ethics according to Scheler),
judgmental acts, norms, rules as well as types of actions and units of
goods, mores and customs, follow the *direction* of the movement of love.
In like manner the areas of aesthetics and the ideas of a *Weltanschauung*
are borne along by the movement of love. By this Scheler maintains
that all historical perspective rests on an order of values and its many
types of comprehension.

   The word "value-ception" (*Wertnehmung*) needs clarification. It can
be understood in the sense of taking something that is already there
(e.g., when values are taken as qualities of things), or in the sense of
taking something which is given *in* the act of taking. In the last sense
and only in it does Scheler understand the term, "value-ception"; for
he says unequivocally that "it is not essential to the act of love, that it
addresses itself in 'response' to a felt or preferred value, but rather that
this act plays a *discovering* role in our grasping of values,"[9] – and that it
alone so plays it – that it displays a movement, as it were, in the *course*
of which ever new and higher values, i.e., yet completely unknown to
the being in question, emerge and come to light. In the course of the

---

[8] *Der Formalismus*, p. 318 f.
[9] *Ibid.*, p. 275.

act of value-ception the resistance factors of willing and handling of practical things and goods are subjected to an emotional value structure, through which all such things must pass. This comes about first in pure attraction and repulsion ("pull" and "push") with regard to the kinds of values things place themselves in. For it is not things that delineate the world of values, but the world of value-essences that determines the knowable being of man, and which world man always has *with* him. Accordingly, man is subjected emotively to a non-formal order of values, irrespective of whether he is in living agreement with it (i.e., ordered love) or not (disordered love). For it is characteristic of Scheler's concept of fate, too, that man neither has to bow fatalistically to his own or a common fate, nor that fate is the product of character or of chance. Rather, man can stand both above and below his fate. "He can be under its spell, so that like a fish in the aquarium he does not even know it to be his fate; or he can, in recognizing it, stand above it."[10] Yet fate remains always determined by that attraction and repulsion, which is in the *ordo amoris*, and it consists in the manifold possibilities within the "scope" of the experience of the emotional coherence of man and world *in* which any real happening first takes place.[11] For the environmental structure of man (*Umwelt*) changes as little as the structure of fate. Only within this structure is the actual surrounding world changed, as well as that whereby man lives in accord with fate. If man's love can be interpreted as an original act of participating in being, i.e., as attraction and repulsion with regard to the essential world of values into which being emerges, its *ordo* consists in the ranking of the values themselves, which are given in the act of value-comprehension. Therefore, the *ordo amoris* of Scheler has an antithetic character. On the one hand it is the source of the dynamic spontaneity of a pure "push and pull" with regard to all matters of love:[12] on the other hand love is the originally structured foundation of value-ception. For the *ordo amoris*, one's emotive attunement or, symbolically, the heart of man, is a "structured counterpart" of the world of values. This is a point that, as is well known, Scheler took from Pascal, and it completely separates value ethics from Kant's practical reason. The a-logical side of man, his emotive attunement, is not a chaos that stands in need of order. Rather, it is ordered in its original manner. And this ordering, as subjective resonance of values,

---

[10] *Schriften aus dem Nachlass*, I, p. 352 f.
[11] *Ibid.*, p. 349 f.
[12] For Scheler hate also is based on love, but this is not the place to go into such a topic. Cf. *Schriften aus dem Nachlass*, I, p. 368.

which in themselves are understood as absolute and relative values, is basically different from logical judgments. But in what does this alogical *ordo* consist in its details? In Pascal's maxim, "The heart has its reasons," what does "its" connote? And what does it mean that the ranks of values and the laws of preference are as lucid as truths of mathematics?[13]

As structured mirror of the value world the *ordo amoris* must reveal itself as a microcosm of value-ranks and in such a way that it has to do only with values and their relations to one another. For, strictly speaking, Scheler distinguishes not two but three relationships within value theory. First, such as obtain only between values; secondly, such as obtain between values and the world of goods that represent themselves in them; and thirdly, the relation between goods and things. Since the *ordo amoris* does not entail any arrangement of goods (and thus also not of things), it can be only an ordered mirroring of values themselves and their relationships.[14] The factual presence of worlds of goods and their transformations is therefore to be kept strictly apart from the *ordo amoris* as such. In this sense there is implied in the *ordo amoris* the apriori of value-ception as well as the fate of the human person. In the third section of our discussion we shall have something more to say about this.

As is known, Scheler distinguished between five value modalities. These are the value categories of the pleasant, the useful, the spectrum from the noble to the common, the values of spirit and of holiness.[15] All values together with their concomitant modalities are divided into higher and lower values, which are grasped *in* the act of preferring or rejecting (not choosing), i.e., in the emotive intuition of preferential evidence. This ordering within values themselves is determined by the following relationships:[16]

---

[13] Cf. *Der Formalismus*, p. 268 f; *Vom Umsturz der Werte*, p. 63.

[14] *Der Formalismus*, p. 36 ff; *Schriften aus dem Nachlass*, I, p. 366.

[15] In *Formalismus* Scheler speaks only of four modalities, in that he understands utilitarian values as consecutive values (p. 124). Yet he regards the same values in "Models and in Leaders" (*Nachlass*, I, p. 268) as a self sufficient modality, so that in this case he speaks not of four but of five types of person. In *Formalismus* he does the same (p. 129, p. 586), and on p. 316 of the same work he indicates at least implicitly the self sufficiency of utilitarian values, although previously he has spoken of but four modalities (p. 125 ff.). What the situation and function of the modality of utilitarian values connotes would need further explanation. Only when these can be satisfactorily eludicated could one and *must* one speak of five modalities of value ethics. Since Scheler constantly held to five types of person (insofar as this is evident in his published works), we can be faithful to his meaning in this instance and speak of five modalities.

[16] *Der Formalismus*, p. 110 ff.

1. A value is higher the more it perdures;
2. A value is higher the less extensive or divisible it is;
3. A value is higher the less it depends on another;
4. A value is higher the more it satisfies;
5. A value is higher the less it is relative to absolute values.

(The lower ranking of a value is to be understood in the converse sense of these statements.)

No matter how one may judge Scheler's thoughts concerning higher and lower ranking of values from numbers 1 through 4, it is the fifth value relation between absolute and relative values on which the alogical apriori of personal value-ception, as founded on love, rests.[17] For it is the absolute values of value ethics which are given in "pure" feeling (love). Here "absolute" means only: *independent of the essence of life and of sensibility.* In this sense Scheler also calls them "unrelated to life." If a value is related to life and sense reality, e.g., the value "pleasant" (animal, man), such a value is apriori lower than a value unrelated to life, e.g., the value of the restful beauty of an art work, that is independent of the conditions of sensuality and thus more lasting. The immediate intuition of this relationship of absolute values to life-values, as given in the *ordo amoris*, in value phenomenology is independent of any judgment which comes about only after an intuitive apprehending. Thus Scheler also states:

Completely independent of "judgment" and "reflection" there is an *immediate* feeling of the "relativity" of a value, for which the variability of the relative value in the mutual constancy of the less "relative" (whether it is a case of variation and constancy in relation to "duration" and "divisibility" or "depth of satisfaction") is indeed a *confirmation* but not a proof. Thus the value of the cognition of truth or of the quiescent beauty of a work of art – utterly independent of the testing of its durability with respect to the "experience of life" which more often may lead us astray from true absolute values than lead us to them – has a *phenomenal disconnectedness* from our transitory vital feelings and thus from our sensual states.[18]

The being-higher of a value lies therefore in the emotionally comprehended *essence* of this value itself. This is evident in the preferring of such a value, that exercises its "pull" prior to any other. There is thus implicit in the *ordo amoris* an apriori value relation by which we prefer an aesthetic value to one related to life, and in the sense of this apriori it is impossible not so to prefer. For even a value deception, in which

[17] *Ibid.*, pp. 117-120.
[18] *Ibid.*, p. 119.

the value "pleasant" would be preferred, is possible only on the basis of an emotionally axiological relationship between life-related and life-unrelated values. The order of values is therefore constant, even in a value deception, and this order is central to the act of true loving.

If the act of love is that act which is basic to all others in the sense that, as pure taking-interest (*Interessenehmen*), it is the presupposition for any taking notice of, any becoming aware of, observing, presenting, willing, judging, etc., then timelessness must be characteristic of it as a pure givenness in man. For only what man adds to it through basic secondary acts is subject to historical variability; but the *kinds* of values, as correlates of acts of love as well as their relationship with one another, are historically invariable. Interdependent with this is the fact that the verb, "to love," logically does not allow of a temporal adverb, so that we could say "I love a person now but not tomorrow." One can see here also that the directions of secondary acts are oriented toward a primary taking-interest-in.[19] For this reason also the acts of preference open up historical variables (e.g., ethos), if for instance certain values of civilization (utilitarian values) are widely preferred to cultural values. The formal laws that exist within the realm of values, as Scheler deals with them, are thus to be understood only as the consequence of the ranks of value contents, as anchored in the *ordo amoris*.[20]

Putting all of this together, we can say that the *ordo amoris* is the subjective side of value ranking, a "microcosmos," which (as emotional) comes before all acts. Its locus is man. This subjective side of man is atemporal, since it is the resonance of the constant (eternal) order between values and determines further knowing and willing for all time. The factual worlds of historical values are to be distinguished from the *ordo amoris*, i.e., the worlds that constitute themselves in the variabilities of preference-acts between constant *kinds* of values. Acts of preferring are based in the primordial act of love whose correlates are values. Therefore, the direction of every preference-act is determined by the direction of the act of love, which in itself can be either correct or not (love and hate).

II

Scheler states that for every object within the realm of values there is a proper place on the scale, to which there corresponds a proper emotive movement and, subtly nuanced, an emotive attunement. Under the

---

[19] *Schriften zur Soziologie und Weltanschauungslehre*, p. 96; Cf. also *Der Formalismus*, p. 108.
[20] *Der Formalismus*, pp. 48, 102.

influence of passion and impulse this originally ordered relational complex of movements towards its value-correlates is overturned.[21] This means that the *ordo amoris* has been violated, so that the direction of love becomes dis-ordered and false (*desordre du coeur*). As mentioned above, Scheler wanted to conduct a detailed investigation of the disorders of the *ordo amoris*. We do not know what he had in mind with such a difficult project. Yet, he designates resentment as *one* of the sources for the overturn of the well ordered direction of love.[22] And here he states quite expressly that the treatise on resentment can be conceived only in such a sense. From this it is already evident that resentment and the *ordo amoris* together pose a fundamental problem, and that the sociological-historical as well as the culturally critical aspect of resentment has to be understood primarily as a *consequence* of the thematic unity of the *ordo amoris* and of resentment. It is thus our task to throw light on both, so that we can come to grips with the disorders of the *ordo amoris* from our insights into resentment, and in this manner grasp also indirectly the *ordo amoris* as the subjective counterpart of any ranking of values.

Scheler also hints at factors other than resentment in the matter of disorders and violations of the *ordo amoris*.[23] These should be mentioned briefly. He speaks of "metaphysical aberrations," when the heart (*Gemüt*) takes a relative value for an absolute one in such a way that the center of a man's personality stands in relation to it as worshiper, as though it were a god. In principle, any finite good can become an idol, as for instance money (as in the case of the miser), a nation (the nationalist), knowledge (the Faust type), physical love (the case of Don Juan). The most common form of such reversal of the orderly direction of love – in our case the displacement of the proper correlate of a religious act – Scheler calls an "absurd obsession" (*Vergaffung*). It manifests itself in the above mentioned sense of the dethronement of deity by a relative and finite value, posited as absolute. A second kind of this infatuation consists in the perverse preference for, or rejection of only relative values. Both kinds of infatuation are involved in resentment. For in all cases there comes to the fore the basic relationship of the tension between (corporal or spiritual) powerlessness and a positive value, which induces in a subject of resentment (individual, group, alliance, race, etc.) the psychic mendacity of deceitful self-elevation.

[21] *Schriften aus dem Nachlass*, I, p. 335 ff., p. 367 f.
[22] *Der Formalismus*, p. 320; *Vom Umsturz der Werte*, p. 63.
[23] *Schriften aus dem Nachlass*, I, pp. 367, 220, 360 ff; *Vom Ewigen im Menschen*, p. 261 ff.

And such self-exaltation consists in tearing down a positive object value, in order to belie one's own powerlessness (e.g., unbelief) by means of a psychically envenomed state of being. At this juncture we need not deal with particular cases and types of resentment, as for example, the case of a cripple, social, political, personal, or situational resentment. What is of import to us at this point is that the *ordo amoris* is clearly being violated in such cases because of the false direction of the act of loving. For the proper value correlate has been displaced, and with this is connected, emotionally, a falsification of values. The relationship between the original order of emotive attunement and its value-correlates is confused in the resentful personality. As to individual traits this complication might well be hereditary, as in certain cases of resentment in cripples. It can derive from the situation in which the resentment-subject finds himself. It can manifest itself in him with unexpected suddenness or it can lie dormant in the individual and the qualitative direction of his personal acts. However it may be, in each case an original structure of order is overturned between the direction of love and its correlates; for, it is basic to value ethics that the scale and ranking of values is apriori-emotional. For this reason Scheler allows for the structure of genuine values even in the resentful man. They glimmer through his illusionary value schema as through a transparent screen. Hence such "organic mendacity" still possesses, as it were, the structure of authentic value relationships, which in extreme cases of resentment becomes subjectively unrecognizable. Still the echelon of value relationships remains "an irremoveable and substantial part of the entire experiential context," which brings about a vital "torture of conflict" between powerlessness and the desire for a positive value.[24] What I call the *topology of the heart* lies at the base of the relationship between the *ordo amoris* and resentment. Just as in an analogous sense one can distort a torus in an infinite variety from insignificant to intensive strength, to the extent of its becoming unrecognizable, without altering the particular topological base-lines, in like manner the *ordo amoris*, i.e., the primal structure of the human heart, is capable of an endless variety of confusions in its value-ceptions. This could be understood in a particular case all the way from a slight impulse of resentment to the desire to destroy the existence of a positive value, or even the complete revaluation of positive values (Nietzsche's "sublime revenge"). This is what Pascal may have envisioned with

[24] *Vom Umsturz der Werte*, p. 51; *ibid.*, p. 65 ff.
[25] *Schriften aus dem Nachlass*, I, p. 371.

his *logique du coeur*, considering the state of mathematical science at the time, when topology as such was not yet developed.

Let us now put this all together in brief. The devaluation acted out by a resentful subject rests on a disorientation of the *ordo amoris*, caused by passion, impulse, and especially by powerlessness (physical or spiritual) which always characterizes the resentful personality. In this way a higher value, which is basically always given him as a higher value, is experienced at a deeper level of the value scale in such a false manner, that he emotionally belies his powerlessness, i.e., the lack of higher value-being. This is based on a disordered direction in loving and leads to an intended (but false) inner satisfaction. In such an emotional process the simple contours of value rankings remain. For positive values, like strength, goodness, holiness, beauty and health, always continue to be encountered by the resentful person, and it is precisely these values that unleash hatred (i.e., false love), which process is rooted in powerlessness. Hence the resentful man always has the selfish desire for artificial satisfaction in his distorted direction of love, and this desire is fulfilled in the devaluating of the positive object-value of another person.

A resentful subject, therefore, finds himself not only in constant self-comparison with others, whose positive worth unleashes the psychic venom of resentfulness, but in addition he always stands in expectation of an emotional *state* (falsely experienced in himself) of higher self-evaluation (*Wertmässigkeit*). For the very sense of his emotive experience consists in feeling an enhancement of his own worth at the expense of the positive value of the object of his resentment; and in extreme cases he might even actually feel he has become the envied object. In all this it is manifest that the whole problem of *feeling-states* (*Zustandsgefühle*), as Scheler deals with them in detail in his *Formalism*, is essentially connected with the *ordo amoris*. For feeling-states are signs of the orderly or disorderly direction of loving. To what extent may we derive all this from Scheler's writings?

It should be mentioned briefly that feeling-states must be kept strictly apart from actually feeling them; and this would mean that a feeling-state can be felt and experienced in a variety of ways. We shall soon return to this important distinction of Scheler's. Feeling-states have in common with values the fact that they all fall into positive and negative ones. In addition to this they have in common with moral values that they – especially as personal feeling-states – attach to acts of preferring or rejecting, like the values "good" or "evil," i.e.,

they accompany realizing acts. This seems to be a point which in its
significance for value ethics seems thus far to have been considered
inadequately. Scheler characterizes feeling-states as the "echoes of
world experience," which we have in loving or hating anything,[26] and
he uses his expression, accompany ("ride on the back of") also in reference to personal feeling-states. Thus, for the further determining of
the *ordo amoris* there is an essential connection: first of all between
value modalities, secondly the acts of preference as based on feeling-
functions, and thirdly feeling-states. If there is order in this context,
the emotional undergirding of the *ordo amoris* must be made to display
itself.

Let us first deal with the feeling-states of pertinent modalities, as
they are to be found in Scheler. To the value modality "pleasant-un-
pleasant" correspond sensual feeling states (as states of sensual feelings),
thus sense-pain, the state of suffering, sensual joy and delight. To the
modality of vital values correspond feeling-states of rising and declining life, feelings of health and illness, being tired or energetic, the
presentiment of death, the feeling of being old, the feeling of rest,
anxiety, or tension. To the modality of spiritual values, there correspond spiritual joy and sorrow as feeling-states (also misunderstood
and mislabeled as psychic feelings). And finally, to values of holiness
correspond feeling-states of blissfulness, of despair, of security, of
qualms of conscience and of peace (also called spiritual feelings). All
these feeling-states fall into positive and negative ones, as e.g., the
feeling of health and of illness. In addition personal feelings of the
third and fourth value modalities have an attachment to the act which
realizes a value, as do moral values of good and evil. It suggests itself,
that in orderly love a positive feeling-state of a person goes along with
"good" and a negative with "evil." That would have to be understood
in the sense that in the acts of preference directed to the value of the
holy, there appears a certain blissfulness as an echo-of-feeling; and
that in an act of the repulsing of values of holiness in favor of another
value, as in the various types of infatuation, a kind of despair, however
slight, announces itself. With this it becomes clear that from a barely
perceptible degree all the way to a full experience of it a feeling-state
can emerge in its givenness. It is of significance for us here, that such a
state attaches only to acts of preferring in a *person*, and that the feeling-
state is not an object of a functioning will. One sees this quite clearly
with feeling-states of sensuousness and of those connected with life

[26] *Schriften zur Soziologie*, p. 332.

values. For in this case the feeling-state is hardly to be regarded as attached to an act of preferring, because the feeling of illness, for instance, is not attached here to a false act of preferring. Hence, feeling-states of personal values are substantially different from those of sensual and vital values, and they betray, as do personal values themselves, a fundamental detachment from the essence of life and sensuality. This means that they are, like these same values, *not* relative to the essence of life. As signs of acts of preferring feeling-states must tell us something about the correct or false direction of love in a man or in a group.

Let us remain with the resenting subject. He evaluates and acts always from powerlessness; and we saw above that he always strives after additional values, in that he detracts from a higher value that challenges emotionally. In the case of the cripple this would be the value of good health, with the socially suppressed it would be that of social right and power. In all of this it is obvious that the resentful man from his position of powerlessness is aiming at a higher value, and also at a more highly valued feeling-state; for, as "good" and "evil" are neither valid as a goal nor as purpose in Scheler's ethics, neither can feeling-states of a person be achieved or sought for. Therefore, the direction of love in resentment is to be regarded as twisted in that a feeling-state of another is supposed to become the goal of inner satisfaction, though in itself a feeling-state must attach only to acts of preferring or rejecting a person. Any means to achieve this inner satisfaction is justified to the resentful man, not only emotional deception and premeditation (as in revenge) but also any possible detracting from the valueness of a higher feeling-state itself.

In like manner beyond resentment we gain some information on the determinants of the *desordre du coeur* which underlies it. For on the basis of what has been said a conative direction in regard to *any* feeling-state, even for one related to the essence of life, i.e., a non-personal one, must violate the *ordo amoris*. Every striving which is directed toward the production of a feeling-state, must come from inner experiences of dissatisfaction, weakness, and powerlessness. This is equally valid for an individual who is intent on higher states that would correspond to inner quietude, as also for any group, that seeks even to build a philosophical substructure under its conative direction towards a feeling-state. Scheler clearly gets at the core of the *desordre du coeur* that brings about false love orientation with regard to feeling-states, when in reference to Eudaimonism he avers:

For wherever in the *more central* and deeper layers of his being man is *dissatisfied*, striving becomes an *attitude*, as it were, of displacing this unpleasant state through intentional conation toward *pleasure*, i.e., with regard to peripheral layers in question, i.e., at the same time the layer of more easily producible feelings. Already the intention of striving toward pleasure itself is to that extent a *sign* of inner unhappiness (despair) or – depending on circumstances – of inner wretchedness or misery, inner dejection or sorrow or a vital feeling, which reveals a trend in the direction of a "deteriorization of life" . . . Also for a whole era exaggerated practical hedonism is always the surest sign of vital degeneration. One can even say that the demand for the means to produce sensual pleasure and to alleviate pain (e.g., narcotics) is as a rule the greater, the more joylessness and a negative determination of life-feelings becomes the inner basic *posture* of a society.[27]

Scheler mentions here a number of negatively determined feeling-states, with which we familiarized ourselves above, e.g., despair (fourth modality), unhappiness and sorrow (third modality), declining life-feelings (second modality), which effect the positively determined feeling-state of pleasure, as that of the first modality functioning as an end of a conative intention. For the sake of making the case more simple, I believe, Scheler refers to Eudaimonism here. For in contrast to other feeling-states of higher value modalities those states of sensible pleasure are more easily producible.[28] For this reason it is natural for a man with negatively conditioned feeling-states, e.g., of sorrow, to be oriented toward the surrogate of sensible pleasure. The above reference cited from *Formalism* is to be regarded as *one* example of the *desordre du coeur*, insofar as Eudaimonism is revealed to be one of its philosophical foundations. As already mentioned, any feeling-state can be theoretically the goal of conative intentions. This is the case when e.g., happiness and faith are the goal and purpose of Sunday church going, when, to put it briefly, church going becomes a kind of technique for the production of inner satisfaction in the sphere of feeling. That a confusing of the *ordo amoris* is here evidenced is not only clear, in that happiness must accompany a true religious act, but also in that faith itself falls under the response reactions of spiritual feelings.[29] In the imperative ethics of traditional ecclesiastical mentors, which Scheler tries to

[27] *Der Formalismus*, p. 357. Here we must make reference to an unclear passage in *Formalismus*. On p. 129 Scheler mentions happiness and despair as feeling states Yet on pp. 354 and 355 he says that they can never be states. And in *Schriften zur Soziologie*, p. 39, he adds to these psychic feelings. This difficult contradiction and its solution lie outside the present scope of our essay, since here we are only occupied with the love direction *toward* a given object, not with an investigation of feeling states as such.
[28] *Der Formalismus*, p. 348 f.
[29] *Ibid.*, p. 129.

pinpoint from the viewpoint of resentment, there exists the "great danger" of taking the duty of going to church, or of cultic rituals, as faith itself. Thus, in place of the duty of faith, which rests only on spiritual acts, one posits symbolic actions and "busy work" as the sense and purpose of the genuine religious spirit.[30]

From the places and cases cited it is obvious that the *desordre du coeur* is present when there is a false direction of the conative intention in such a way that a feeling-state, which as a sign can only accompany acts of preferring, is sought for as a goal of gratification. Or expressed in another way, in the *ordo amoris* a feeling-state cannot be intended; rather, the latter is only a spontaneous echo of the act of preferring or rejecting values. Thus, a feeling-state must of itself accompany every act of preferring. *One* case of disordered love, therefore, is to be seen in the conscious objectifying and intending of feeling-states, which by their nature only accompany acts of preferring values given in pure experience. If a feeling-state of another is objectified out of envy, malice, revenge, spite, jealousy – the initial forms of resentment – a false application of acts of willing and striving for the emotional being of another person is in evidence. If in the absence of any objectification in preferential value-acts a feeling-state, as accompanying them, is *experienced*, a person is at the least on the way toward the *ordo amoris*. In all this one can observe that not only a positive feeling-state gives evidence of the *ordo amoris*, but also a negative feeling-state can be experienced as such in its purity. Thus pain, as a feeling-state of the first modality, can witness to the *ordo amoris*, namely when it permeates the entire personal sphere, in that a person gives himself wholly up to pain and suffering. The redemption from pain and suffering does not consist in their conscious lessening or alleviation but rather "in the art of suffering in right manner," i.e., to suffer freely in one's soul (to "take the cross willingly upon oneself").[31] When Scheler refers here to cheerful suffering, we have to be mindful of the following circumstances. As stated at the start, as pure attraction and repulsion with reference to the scale of valueness of all beings, love and hatred are original acts on which all other acts build (in which case even hatred is based on the pure taking-interest-in of love). On the other hand feeling-states exist as the echoes of world experience, which themselves are only nameable (*konstatierbar*), i.e., are contents,[32] but are differently

[30] *Der Formalismus*, pp. 235, 241, footnote 1.
[31] *Der Formalismus*, p. 359.
[32] *Ibid.*, p. 270 ff.

colored (historically) through feeling-functions. In this context Scheler mentions the feeling-state of suffering,[33] as it permits itself to be experienced in dedication, enduring, tolerating, enjoying, etc. One must accordingly broaden the emotional context of values, acts of preferring and feeling-states relating to functions of feeling. We thereby obtain the following schema:

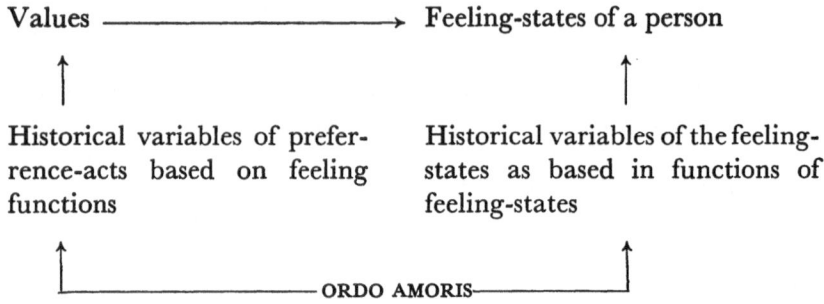

It is to be understood that in this schema we are dealing with a struc- ture of an order, not with a presentation of a time sequence of acts, evaluation, and feelings. Likewise, the direction of the arrows in our picture are to be taken only as provisional. Thus, for example, one could let the arrows between the *ordo amoris* and values go in the opposite direction, especially if we understand the *ordo amoris* as the subjective side of the ordering of values, as we have done. Yet values, as act-correlates, are also the objective side of the *ordo amoris*, as the schema should indicate. Similar things may be valid for the direc- tionality between feeling-states and *ordo amoris*. In no case, however, is the ordered direction between value and feeling-states of a person to be altered (as discussed above).

In this picture two different emotional layers manifest themselves. In the first we have the constant types of values and their rankings. Then in the same layer we have the contents of feeling-states which also may not be altered. The second layer is that of variables of preference- acts and feeling-functions that have temporal significance, since through them the data of the first layer are comprehended and experi- enced in a temporal dimension. *Between* both layers there lie for Scheler the possibilities of various moralities, ethical systems, and re- ligious redemptive doctrines, that undergird philosophically or theo- logically the variability of the direction of acts and feelings and give

---

[33] *Schriften zur Soziologie*, pp. 37, 53 f.

the "spiritual personality" freedom to interpret feelings and values both in a religious manner (salvation theory) and philosophically (ethics).

It is clear that the *ordo amoris* of man not only amounts to an emotional resonance of the correct amatory direction and therewith to the correct act of preferring, both of which express at every moment the personality's fundamental amatory direction, but that also the proper way to experience feeling-states on the occasion of respective acts of preference or of non-preference is constitutive of the *ordo amoris*. While preference-acts are properly determined by ranking of values, the normal direction of feeling-functions to feeling-states is, as we have explained, not determined by their ranks which are parallel to the ranks of value. For the feeling-states of a person accompany a respective preference-act. And thus the proper way to experience a personal feeling-state lies in its free unfolding and in letting it come about by itself, not to force it to happen nor to try to alleviate it by means of psychic techniques, and certainly not by artificial means (such as narcotics). A *minimum* of feeling-functions allows a feeling-state to conjure up personal being *as* its self in toto. The essence of the inner law of the *ordo amoris* does not consist in the proper act of avaluation or value-ception alone, but beyond this in the *readiness* of the person, freely to accept the emotive echo of value-ception both as a positively or negatively determined feeling. It is in this sense, we believe, that Pascal could behold in Christ the example *par excellence* of the *ordo amoris*.[34]

Resentment contradicts the alleged ordering structure in the following obvious ways. Resentment is always the result of a conscious state of powerlessness, which, however, is not taken by the person as such, but *to* which resentful man relates, while he simultaneously and continually represses release of resentment while comparing himself with other persons. Not only a falsified value comparison with the object person of another value level but also a comparison of feeling-states is obvious in resentment. For who could not see, e.g., in the resentment of certain cripples, in the resentment of the generation gap between young and old, a comparison of states between consciously experienced physical powerlessness or the emergent feelings of life and health? Quite obviously, the relation to a feeling-state in a group or circle, which already senses its condition as a fate and considers it a matter of revenge, is apparent in forms of race-resentment. Resentful

[34] Cf. *Der Formalismus*, p. 269.

man, therefore, not only detracts from the value level of the other person through psychic self-envenoming, but he seeks at the same time to alleviate his feeling-state through feeling functions and to falsify that of the object person. Thus not alone in the falsification of values and in value-deception on the basis of disordered acts of preference, but also in the continually conscious inhibition to *be* his feeling-state, to let it be fulfilled while at the same time objectifying the feeling-state of the object person in order to attain (falsely) inner satisfaction, we see the *desordre du coeur* lying at the base of resentment.

It is difficult to grasp the *ordo amoris*. For since it belongs essentially to the sphere of *person*, it is, as the person himself, incapable of becoming an object. It would thus be wholly wrong to understand it in some way as a value monad or as a substance. The *ordo amoris* withdraws itself from thought, because not the ordered structure itself but only the latter's grasping and obtaining, viz., through the deception and overturning of this order, which lurks in resentment, is historical.[35] Therefore, there exists one possibility of laying bare the *ordo amoris* in that one investigates its aberrations in resentment by describing them. May our reflections, which constitute a partial investigation of this possibility, contribute to this attempt.

III

The above discussion leads us to an interpretation of the passage quoted at the beginning of this paper from Scheler's treatise, *Ordo Amoris*, which we consider essential for his entire (published) writings.

We already saw that we can understand the *ordo amoris* in such a way, that it furnishes subjective ground rules of values in an analogous sense to the topology of a physical body. Thus it is comparable to a "crystal formation." These basic lines of our emotive attunement determine for Scheler the what and the how of knowledge and of willing. For the order of values as such is what man carries with him like a "housing,"[36] and since feeling-states of a person, like moral values, always attach to the act that realizes a value, all types of feeling-states are, as it were, the echoes that resound in this housing. The direction of love and hatred thus circumscribe the scope of feeling-states.[37] Thus, as a person, man is fated. For his history, world views and norms, his forms of ethos and his global experience remain within this housing, and everything that

[35] Cf. also *Vom Umsturz der Werte*, p. 63.
[36] *Schriften aus dem Nachlass*, p. 348.
[37] *Ibid.*, p. 373.

can happen to man can do so to *him* only, as dweller within this value-frame. Whatever he experiences in this housing by virtue of the variability of acts of preference and feeling remains determinative of his further knowledge and willing. While everything sensible and all feeling-states related to life modalities are an inescapable fate of *everything* alive,[38] the frame of feeling-states not relative to life is specifically human fate as the fate of the *person*. For it is the personal unity, independent of generic organization of the biological creature, man, in which acts thus appear as given, so that there opens up in this givenness a "split," in which values are manifest that are above the essence of life and whose rank-structure determines what is experienced and created in time.[39] The possible worlds of an individual person or group are ruled by ontological directions of love and value preference, by the "relief" of pure attraction and repulsion through the order of the heart. These basic lines are experienced in the pertinent correlativity between value-ception and feeling-states, and thus they are the basic emotive tenor of the person.

From this there accrues not only historical but also ontological significance to models of personhood, as Scheler takes them, i.e., person types such as the saint, genius, hero, the leading mind of a civilization, and the *bon vivant*. As personal embodiments of value rankings, whose subjective resonance rests in the *ordo amoris*, they are the "polar star" of the human race.[40] This must not be viewed as though model personages arose historically only through acts of preference or were deduced from historical types. For as "schemata" that express basic lines of all human love and of its basic values in personal form, they *possess* us, before we can choose among them.[41] And thus Scheler understands the sense and essence of history as that of model personages (and their opposites), which, taken in themselves, are ideal types grounded in the *ordo amoris*. Yet in time they must submit to the bloody baptism of experience and history, in order to enter concretely into it.

When Scheler states that the *ordo amoris* is the hidden wellspring of that which issues from man, its ontological sense lies in the additional phrase "even more" used in the passage. For when models already "possess" us, they are that which posits itself about men, his moral surroundings, as effected through his *ordo amoris*. And these models and

---

[38] *Schriften zur Soziologie*, p. 38.
[39] *Der Formalismus*, p. 302.
[40] *Schriften aus dem Nachlass*, p. 269.
[41] *Ibid.*, p. 268; cf. also *Philosophische Weltanschauung*, Dalp, vol. 301, p. 33.

the higher value ranks are in time his fate, since they incarnate man's being into its historicity and determine the being of his person. This emergence of the person happens through the aforementioned split. For since man as biological genus has lower value-classes more or less in common with animals, in the global openness of the sphere of the person the value-experience of being, as detached from the essence of life, gives of itself. At its base lies the emerging and passing away of being in participation and sharing with another being, a "breaching of the limits of one's own being," for which Scheler sees "no other name" than love (in the most formal sense of the word).[42]

If the value factor of being, as detached from life-experiences, opens up in the personal sphere, as yet nothing is said concerning the Being of beings. In this sense it does not seem reasonable to us to disregard ontologically the anthropological starting point of Scheler. Rather, the situation seems to be such that the attempt to gain sight of Being from the basis of the *ordo amoris* sets a task for the *ontology of human existence.* For if we regard Scheler's fundamental position as correct, if indeed love and value-ception precede knowledge, the question arises whether it is precisely this original emotive value-taking, as pure taking-in-terest-in, that makes the Being of beings withdraw from us. Or rather, whether Being announces itself through its being Value as the sense or a sense of Being, to be found in the human person, viz., whether the value world of a person is *access* to Being. These questions need not be regarded only as anthropological. For if the basic direction of loving and hating determines the scope of knowledge and human effectiveness as well as the possible world of a person "ontically," as Scheler states, then ontological meaning does actually accrue to the *sphere* of the person, an expression he liked to use. This should be so understood that the types of values detached from life, viz., the values of right and wrong, the values of the cognition of truth, the value of beauty, the value of the holy, but also moral values, all as personal values, make Being appear within the split. Thus the word "holy," and its model-types present the high point of this realm of values. In this the question of God is not yet touched upon. Here one could decide only to a certain extent whether especially the value "holy" as related to absolute objects (in opposition to all other values), abets more than anything else the ontological moment in the personal sphere by what was thought of, or passed for, the absolute at various times, or how Being allowed itself thus to become thematic for thought and even predominate in in-

[42] *Ibid.*, p. 40.

tellectual history. In this context no small significance acrues to a person's feeling-states as "echoes" of world experience. It is not by chance that Scheler considered anxiety as a *feeling-state*, i.e., of a state of vital values, and regarded it together with all spiritual vitalities as the source of civilization. [43]

These questions indicate that the *ordo amoris* is to be ontologically thematized; and Scheler may well have had this in mind, but he could not treat of it anymore. When he avers that man's experience of reality is *resistant* experience, he means that in a living being without any form or content, i.e., without consciousness of color, tone, or body, without the forms of space and time, etc., there remains an irreducible impress of reality, a bare factor of resistance, being of an undetermined character, an empty What. When he states that this resistance is *"there"* even *"before"* all perception and thinking, we stand within the framework of the *ordo amoris* before the question of how the spontaneous "choosing mechanism" of the *ordo amoris* selects from the pure "there" of resistant Being a world of values (with regard to the human person) as an "island" emerging from the "sea" of Being. If one regards the value world as a calculus with which one can get at the sense of being only to some extent or not at all, we may counter this opinion with Scheler's tenet *that* Being indeed does allow an essential world of values to break forth from the pure "there" of world resistance through the person's *ordo amoris*. And last but not least we are reminded that Scheler understands the person *only* as correlated with world, and world *only* as correlated with the person. At this juncture it seems that there is some similarity in thinking between this and the Thereness of being-in-the-world.

It is customary to place Scheler in the history of contemporary philosophy between Husserl and Heidegger. No matter how one may regard Scheler's ideas on phenomenology, perhaps as a transition between Husserl and Heidegger, such categorizing makes sense only if one can establish a relationship between the non-objectifiable person and Heidegger's "being-there." That Scheler himself stood close to the thought of *Sein und Zeit* is witnessed by the fact that at the time of its first printing he was one of the few, if not the only one, who immediately realized the scope of the question in *Sein und Zeit*, and until his

---

[43] *Schriften aus dem Nachlass*, p. 308; *Schriften zur Soziologie*, p. 39.
[44] *Die Wissensformen und die Gesellschaft*, p. 363; cf. also *Vom Ewigen im Menschen*, p. 215; *Der Formalismus*, p. 155.

death he was in dialogue with it.[45] Still, a clarification of the problem of the person vis-à-vis "being-there" becomes possible only after the marginal remarks of Scheler in his own copy of *SZ* will be edited in full in the *Collected Works of Max Scheler*. These notes should contribute further in an essential way to our intuiting of man's *ordo amoris*.[46]

transl. F. J. S.

[45] At this point I may be allowed to make reference to a letter from M. Heidegger, August 6, 1964, in which he informed me of his last three day visit with Max Scheler in Cologne in the winter semester of 1927/28. I am deeply grateful to Heidegger for this reference.

[46] Just before this essay went to press Prof. Maria Scheler, previously editor of the *Gesammelte Werke* put Scheler's own copy of SZ at the present author's disposal in Munich. Since then this material has been expressly handled in *Person und Dasein* (The Hague, 1969) by the present author. *Editor's note*: M. S. Frings' book, *Person und Dasein, Zur Frage der Ontologie des Wertseins*, Phaenomenologica 32, has been reviewed by the editor/translator in a forthcoming issue of *Philosophy and Phenomenological Research*. F. J. S.

Source: *Zeitschrift für philosophische Forschung*, Bd. XX, 1966, Hft. 1, pp. 57-76, "Der Ordo Amoris bei Max Scheler, seine Beziehungen zur materialen Wertethik und zum Ressentimentbegriff."

Don Ihde

Sense and Sensuality

Were you to look at the best sellers' list this week you would find at least three or four of the leading non-fiction books dealing with sex. In the fiction category sex is so pervasive that to find a book without at least one vivid scene of coupling would be a considerable task. But between the non-fiction and the fiction books there exists a gap. The non-fiction entries seem to be dominated by a "technology of sex": techniques, physiology (or "plumbing"), all described in clinical terms as if some observant Martian were viewing us as a curious race of "interesting mating animals" with an insatiable passion for copulation.

In the fiction category there is a certain detachment as well – but with a different emphasis. The copulators in this category are mainly self-enclosed subjects rich only in fantasy life. Indeed, one such book consists of the recitation of the wandering thoughts, daydreams and memories of a woman undergoing seven minutes of intercourse.

It occurred to me that this plethora of often asexual books about sex did indeed say something about our post-Victorian preoccupation. The pattern seems to be that either we regard sexuality as the mechanical action of an object-like body which needs to be tuned up 'now and then to perform well, or we regard sexual relations as the excuse for subjective fantasies. In short, the glut of literature can be seen as a form of *institutionalized cartesianism*. We are acting out the dictums concerning man as body-machine whose interior, private self is inhabited by "mental events," the vestigial remains of the cartesian thinking substance. And we are troubled by the connections, usually vague associations, between ourselves as objectified bodies and our hermetically sealed inner experience.

A very serious instance of the contemporary tendency to take cartesianism literally exists in a mental state, schizoid in form, called "depersonalization" by J. M. Heaton:

In depersonalization the patient feels compelled to observe his own actions like a spectator. This has a destructive effect on the immediacy of his experience and almost any area of it may be affected. Thus every thought and emotion may be taken to bits and the patient may feel like an automaton without a will of his own.

His spatial experience may be altered so that the world seems perfectly still, like a postcard. Things may seem remote and often there is a general impression of flatness. Time is altered so that there is no longer a sense of becoming and all seems timeless, unchanging and hopeless.

The patient's experience of his body is often changed so that it may feel strange and as if it was not his own; often the patient feels he is existing mainly in his own head, usually somewhere behind his eyes through which he has to peer out. ("Depersonalization and the Development of Visual Literacy," J. M. Heaton, *Confinia Psychiatrica*, 11, 1968, p. 181.)

This experience is cartesian even to the final location of the observer consciousness somewhere in the head behind the eyes – perhaps in the pituitary gland thought to be the connecting place between the body and the soul.

I could be tempted at this point to begin an extended and direct attack upon this version of cartesianism and its inadequacies, particularly in regard to experience – but instead I shall merely note it as a background problem which pervades our beliefs and language. Instead of an attack, I want to begin a re-tracing of a path back towards a careful examination of the richness and complexity of concrete experience and this is the context of a phenomenology of eros. Even in the realm of the erotic the cartesian dissociation of me from my body and from my acts has occurred.

My strategy will be one taken by phenomenological and existential philosophers. I wish to affirm that *I am my body* – but that this embodiment which I am and which I enact is not that of cartesian machine, but of a person of living flesh. After a brief comment about such a phenomenology I wish to outline (a) some characteristics of the experience of *touch* as they relate to structural features of experience; (b) then move the discussion to some of our rites of invitation which encircle our "lived space" and which relate to sexual spatial constitution; and (c) finally, I shall indicate why, from the preceding considerations, the sexual act even in relation to touch and distance is privileged in certain respects.

Phenomenology at its inception began both as an anti-cartesian polemic and as an attempt to recover the clues for a full appreciation of experience. Husserl, particularly in the *Crisis*, began to speak of a *life-*

world and of the embodied person, and certainly the existential pheno-
menologies have continued to investigate the implications of this recovery of experience, particularly Gabriel Marcel and Maurice Merleau-Ponty. In this philosophical attempt to recover the full sense of
what it means to be a creature of flesh the central feature of experience
was termed *intentionality*. This term, as used by phenomenologists,
characterizes the structural features of experience (also called subjectivity or consciousness).

In briefest form intentionality unites sensing and signification, perception and meaning. Phenomenology rejects the notion of some undetermined or uninterpreted datum below the level of signification in
our experience of the world. It turns instead to naive experience as its
primary field of evidence. But from this primary field it seeks to the-
matize and extract those features of man's relationship with his surrounding world, those structural characteristics which allow an understanding of the mundanity in which we all operate. Phenomenology
finds naive experience to be already rich and full – and the phenomenologist's task becomes the explication of that richness and the clarification of that significance in relation to the structural features of experience.

In the context of a phenomenology of eros, I want to begin my
contribution by descriptively opening some of the aspects of *sense and
sensuality* first in the area of touch perception. The turn to perception
first, I would point out, is also a typically phenomenological move.
The "primacy of perception" asserted by both Husserl and Merleau-
Ponty is an always present *focus* of experience – we are focally perceptu-
ally engaged with the world.

The choice of touch phenomena as a prologue to the phenomenology
of eros should be obviously appropriate. Touching in the most intimate
sense is clearly sexual. But even as I begin two problems pose themselves: on the one hand I cannot re-trace and outline all the conceptual
machinery which I presuppose. The "tribal language" of phenomenology (epoché, the phenomenological reductions, eidetic intuition,
etc.) would draw us too far afield for a short essay – so I deliberately
take the risk of certain shortcuts which may result in misunderstan-
dings. In my descriptions I shall deliberately take examples from ordinary experiences to point up what I am locating. Often from only
one or two examples I will appear to generalize about some structural
feature of experience. The danger which this shortcut faces is that I
may leave the impression that phenomenology is a type of common

sense philosophy which uses induction as its method – and that is not the case. I would therefore point out that my examples are carefully selected to illustrate points which are and should be arrived at only after widely ranging "perspective variations" from which "invariants" are "eidetically intuited." I can only hope to be suggestive for the non-phenomenologist and hope that the phenomenologists will forgive the shortcuts.

On the other hand the selection of touch as a sensory variable risks a very unphenomenological result. Our naive experience is first and foremost global – in our involvements with the surrounding world we normally do not select out a "sense" and, phenomenologically, the so-called five senses are already a kind of reflective abstraction. Nevertheless, I can clearly focus upon touch within this global experience – but I must be aware that I am already doing a first order reflection. My reason for beginning in this way is that by simplifying the examination in considering only one sensory variable a certain clarity may emerge more easily than might be the case otherwise. Thus I choose touch for the sake of initial simplicity, because of its clearly important role in sexuality, and because it quickly illustrates the clues to my fleshly embodiment in the world.

But as I begin to describe what occurs in touching I find myself immediately faced with the traditions of our cartesian heritage. The term, body, suggests an object, something objectified. And if I try to think of this objectification in perceptual and ordinary terms it might perhaps be thought that I-as-body am co-equal to an object, perhaps co-extensive with the visual outlines we perceive. Note that this *idea* we get of ourselves may well be drawn more from our experiences of others than it is of ourselves. We see him in profile – but we see only a very limited aspect of ourselves. We then may leap to the thought that I am "like" him – as a separate object-body. That more than this is involved even with the perception of others I need not point out here – but in this cartesian context I further apply to myself the notion that just as he appears to me without the full perspective of his "inner sense" so even more strongly may I think myself an object-body. Phenomenologically it makes all the difference *how* I gain whatever knowledge I have. In the case being described I would only point out that I never get a full detached perspective upon myself in the same way that I do upon the other. And secondly, I experience a full "inner sense" of myself in a way which is different from the apperceptions I gain of the other. The knowledge, the *idea*, I get of myself as "like" the other is not co-extensive with the "feel" I have myself – nevertheless, the idea of myself as object may lead me to conclude certain things about what I *must* experience almost in spite of what I *do* experience.

This applies to the experience of touch. I might conclude, for example, that because my skin is presumably the place where "touch sensations" occur this must be the place where I actually feel – and this would mean that touch is not a distance sense. But in my role now as phenomenologist I "bracket" all such assumptions, inferences, and theories of causation in order to first turn to naive experience as it gives itself out. I must first locate myself in these acts.

First example: I find very quickly that my experience of touch does not at all remain within the confines of my body-as-outlined object, but I find in certain cases, indeed many, that I can feel and touch at a distance. The classic example is the blind man and his cane. He feels the sidewalk – at the end of his cane. Take a pencil and run it along the desk or the blackboard and take very careful note of where the feeling is (close your eyes if necessary) and you soon discover that the feeling is at the end of the pencil – or, better put, you feel the desk where the contact actually is – at the junction of pencil-desk. My sense of touch exceeds my bodily outline.

Even more remarkable is the skill developed by the physician. In the pencil example the surfaces were hard and perhaps surface and texture is what you first noted – but the doctor as he thumps upon you does not feel the surface so much as he feels your interior. The size of your heart, the configuration of your liver, the arrangement of organs or nodules are felt inside you. His touch penetrates the surface of your body to the interior.

The cartesian, of course, has elaborate ways of attempting to save his presuppositions when confronted with such odd, but ordinary experiences. His explanatory acrobatics attempt to continue his notion of sensation as "inside" and that which "really happens" as an unexperienced "outside." But in my present role I shall no longer argue with him, I merely turn to what the phenomenologist derives from these and similar experiences. These experiences illustrate the first feature of *intentionality*. Experience, consciousness, is consciousness *of* x. Experience refers, points to, is directed into the world. Its first movement is not "inside" but it is outside in the world. I feel the desk, the doctor feels the shape of my heart, the first reference of intentionality is the *Other*.

So much is this the case that in our ordinary involvements with the world we hardly take notice of our feelings in any explicit sense at all. We are immersed in our projects. The blind man's feel for the sidewalk, although he is clearly aware of things that we overlook altogether, is secondary to his project of getting to where he is going. Return to the pencil on the desk example. In this case my direction located the project on the feeling itself – my question set the context. What "stood out" was the feeling of pencil-desk. We felt the desk as the doctor felt the size of my heart, and so on. Our experiences were focused. But a focus is a selection, an emphasis, a theme within the totality of experience. It is not all that is going on.

If I now ask you to repeat the experiment of the pencil on the desk and ask you what else you feel in addition to the first thing that stood out you would soon notice that you also feel – at one and the same time – a vague pressure on your fingers as the pencil traces its route on the desk. And by conscious effort you can even begin to make this finger-pencil aspect of the experience stand out. But when it does, the pencil-desk aspect tends to fade and vice-versa. While both aspects are present one tends to stand in the center of the focus while the other fades to a fringe awareness.

Once again from these simple experiences I generalize to another structural feature of intentionality, the feature I call the *core/fringe structure*. Simply put, in ordinary consciousness we have a core of attention around which but at the same time are arranged a number of fringe experiences. I can vary this focus at will (or I may be called upon to switch its direction by some distraction) but in the normal situation, what is core "stands out" while what is fringe is only vaguely present.

Now let us begin to vary our examples farther. I lie down on a soft couch and begin to read. As I become absorbed in my reading the touch relationship between the couch and my backside begins to fade so far to the fringes of my consciousness that I find that I seem to be almost floating. The more usual sense of semi-weightiness which I feel when I walk or even when I sit down is replaced by this semi-weightlessness. As phenomenologist such an occurrence may become interesting and I switch my attention to it. Unlike the pencil example in which I was able to detect very precise differences in the object I was feeling through it, I find that the cloud-like couch-me experience is so vague that not even any clear distinction between me and where I end and couch is capable of being made. Inner and outer, subject and object are here not at all clear and distinct.

At the couch-me end of the continuum there lies a vagueness which perhaps disturbs some philosophers. I quickly rush to the other end of the continuum of touch phenomena and recall the experience of the safecracker: his deft fingers, sandpapered to supersensitiveity, feel the fall of the tumblers inside. And the quality control man in a paper plant who, by letting the rapidly rolling ribbon of paper glide between his fingers, is able to detect even the slightest variance or imperfection. Surely here is clarity. Note, now, two things. First, phenomenologically, experientially speaking, any reference of attention is a positive phenomenon. I can concentrate and focus upon vagueness just as well as I can upon precision. Furthermore, there is no experiential reason to preclude one phenomenon being any more important than the other. Second, in the continuum of vagueness to precision of which our touch is capable of locating, I find that some areas of my bodily experience are more sensitive than others. In some respects I and world tend to merge, in others I am embodied precision through which I operate upon the world.

To this point and within the limits of ordinary experience we have noted that: (a) touching is intentional in the sense that its reference is the other. It is directed towards the other in the world; (b) within the totality of experience and within touch there is usually a focus which stands in inverse relation to those aspects of the touch experience which stand on the fringe. In this case, that which stands out is clearly present to my experience, while that which is on the fringe is barely noticed; (c) but there is also a continuum of feeling within the range of touch such that in some cases the other seems to merge with me in a vagueness which blurs the distinctions between subject and object – although at the other end of the continuum there may be a greatly refined precision in which I am able to detect the minutest differences of that which stands over against me.

Now, however, it is time to begin noting unique features. The central such feature of touch has long been noted as touch-touched: whenever I touch something I am also touched. Obviously, great differences occur between my initiation of a touch occurrence and one in which I am suddenly touched by another – but in both instances the contact is focalized in touch-touched. I touch the cold wall, and it touches me with its coldness. I touch the slippery vinyl of the contemporary art exhibit, and it touches me with its invented sensuality. I touch the warm flesh of my beloved and her flesh touches me with warmth. *This* is the lack of distance that is so often attributed to touch –

it is not distance which makes the difference, it is the latent and *implied intimacy* of touch in being touched in a necessary mutuality which makes the difference. Touch in this respect is not quite like those dimensions of experience such as sight or hearing – not every time that I see am I seen, nor am I spoken to everytime I speak although I do hear my own voice when I speak) – but every time I touch, I am also touched. There is a necessary mutual intimacy to touch.

We could have noticed this feature in any of the first examples: in touching the desk I am touched; in lying on the couch the pleasant semi-weightlessness creates its touch-ambiguity; in touching the paper it touches me with its smoothness or roughness. All of this lies within the focal possibility of touch. But I have already hinted that in addition to focal possibilities, there is a field of fringe features as well. Touch is also *total*, a plenum never empty of an overall intimate contact with the world. This field of touch is constant even though we are seldom explicitly aware of it.

There are at least two approaches to noting this feature of touch experience. One is by varying our focus within the touch field through a whole series of examples which gradually point up how broad and total our contact with the world through touch is. I note that I do have a fleeting or vague awareness of the fit inside my shoes and that when I walk I am aware at the fringes that the floor feels different when it is carpeted and when it is not. I note that I do have a marginal consciousness of the touch of my collar and the feel of my coat upon my back. And as I explore these fringe perceptions I may even begin to note that there is a very faint feel of the air against my face which is usually a bit cooler than the feel of the rest of my body. And I may note that in some sense all these factors are present within the same period of time although I rarely pay attention to any or all of these features. There is a field of touch within which my specific touch attention is focused upon only some small and highly selected aspect.

But these field characteristics may also be discovered at a stroke, in a sudden *field state*. I plunge, as is my particular passion, into the trout brook in Vermont late in the evening. The almost icy water surrounds me – and I am acutely and instantaneously aware of that cold environment which embraces me. A little later when I emerge, the relative warmth of the evening air gives me a feeling of exhilaration throughout my whole body. Still later as I sit before the outdoor fire I am also aware of the crispily warm feeling of my frontside while my backside begins again to feel the creeping cold of the evening. Finally, relaxed and no longer concerned with the projects of the day, I slip into bed and pull the soft sleeping bag around me and the whole of my bodily being is warm and comfortable.

In these examples the field states of touch become vividly clear. And in these experiences, when the whole of my touch field touches and is touched by the surrounding world, I realize how intimate is the I-world relation in touch. Through touch, I am constantly "in touch" with that which surrounds me. But also in these states it is difficult to say just where I end and world begins. All the specific touches found in focal attentiveness are never separate from total Touch as the constant field in which I live.

I deliberately selected enjoyable field states to first illustrate the touch plenum – however, there are also negative states which could be noted. As a boy on the farm, I recall that one of the most unpleasant such states occurred during the harvesting of oats. The chaff somehow was able, through the course of the day, to permeate my clothing and eventually turn the whole of my body into an itching, hot-sweaty misery. Wanting to escape this unbearable situation became my primary desire through the day and my greatest hope was that the barrel which contained water for a shower had been filled that morning. To rid oneself of a negative field bespeaks the potential horror of being immersed in a world not of one's choosing.

To this point I have tried to show that all touch displays a structural intimacy between me and the world. I am in constant touch with that surrounding world although I am so familiar with my usual surroundings that only on certain occasions – such as those which dramatically demonstrate field states – do I become aware of this total immersion in the touch field. More ordinarily, I am at best concerned with focused modalities of touch, the highlights and most obvious touch experiences which are located by my hands and feet – and as a professor, my seat. I would like to say that the touch world is like being in the presence of an ongoing symphony of many sound modalities – but that I usually pay attention only to the most obvious melodies.

I have noted that although touch seems to require some form of contact with its referential other, this contact in touching finds me touched. Its limits do not necessarily remain within the limits of my bodily outline. It is here that we arrive at the limit and artificiality of dealing with touch alone rather than global experience. I see textures, for example, and apperceptively "feel" them even though I don't literally touch them. Within global experience touch is thoroughly present.

The language of intimacy in touch has also been used deliberately. The structural and unavoidable intimacy of touch is clearly closely related to any serious attempt to deal with a phenomenology of eros. The arrangement or organization of the erotic is a complication upon all and every feature of touch perception. The relationship between touch and eros has long been partly understood in the suggestive metaphor: touch is the *language* of love. In some respects this is a good metaphor. If touch is the language of love, then it may be seen that this language has both surface and depth "grammars." It may be noted that language is extraordinarily complex, rich, and flexible just as are the possibilities of sexual touching. The "screaming curse" of a rape is a very different "word" from the gentle suggestion of invitation issued in the night my beloved makes in the barely perceptible movement. But both remain within the same language of touch and thereby the ambiguities are already problematically announced. The rape is not without its intimacy any more than the lovers' couplings are without the possibilities of aggression and miscommunication.

Such a higher order phenomenon ought to give pause even to the most imaginative investigator so far as finding invariants is concerned. But the circle between touch and sexuality must begin to be closed. Thus by first making an initial connection between the richness of touch possibilities and the sexual orchestrations which are possible upon this basis, I move to the second dimension of the arrangement of touch distances. Intimacy in sexual touching is related to the whole of touch intimacy in the specific rites of invitation.

Between our touch and being touched lies the organization or the "constitution" of space in its concrete experiential form.

I choose to remain within the general confines of touch experience as I approach the subject of the erotic. I want to deal with a certain series of relationships with distance and invitation. A new theme would be entered if I turned from my previous emphasis upon my touching the other and changed the emphasis to that which touches me. In our life-world, after all, on many occasions it is the other who initiates or occasions the touch. These experiences can be disturbing.

I am walking in my dimly lit cellar seeking the right bottle of wine for the dinner and as I move into a corner the barely felt tingling of a spider web, followed by the delicate pattern of feet on my cheek brings a shocking touch – I am touched by the unwanted otherness of the spider and his house and I quickly respond by brushing him off. I rid myself of this intrusion and place this creature at a proper distance from me. The spider has violated the sense of distance I wish to maintain. Between me and others, particularly strange or unfamiliar others, I pose an invisible barrier which is to be penetrated only by invitation or permission. I am not very unlike the lizard sunning himself on the wall who will ignore me until I come within that defense space – then he will dart into his hole.

The lifespace around is constituted space. It is filled with unspoken but definite senses of meaningful possibilities. And because touch is structurally intimate it is also well ordered according to these patterns of invitation or rejection, of rituals of acceptance or rites of warning.

I am an athlete, and leaving the locker I reach out and pat my teammate on the rump in a gesture of support and encouragement. But while I might never think of it, I know that I wouldn't do the same to my colleague as he prepares to enter the lecture hall. Or, I might give a friendly and light blow to the chest of a long forgotten high school friend – but I would hardly do the same to my secretary when she came back from vacation.

Our touching is intertwined with unseen but pregiven meanings. We carry with us an easily felt warning – do not penetrate too quickly or too closely unless you are invited. Of course, there are indefinitely many ways to structure our touching of one another – but these ways are constituted nevertheless. We may form a sensitivity group which evolves new understandings and within the protective confines of "our group" we may get to the point of taking off our clothes and huddling together – but only after having gone through the requisite rituals and never yet do we perform the same acts spontaneously with the saleslady in the department store. Our concrete existence is an ordered world in which we feel, more than know, how we may or may not touch.

Everything has its place in lifespace and he who does not have the implicit understanding of the concrete gesturings may cause misunderstanding or create violence. We introduce each other with the touch of two hands. But we talk to one another at a certain comfortable distance. An old Eastern European friend always made me feel uncomfortable as the thrust his face to within inches of mine to talk. And I would back away until the two or three feet of Western talking distance was re-established. To me he seemed aggressive and threatening; to him I seemed to be rejecting his friendly desire to converse as comrade.

Sexuality, too, is organized and constituted within a context of implicitly understood arrangements of distance, invitation, and touch. This even in the most extreme examples. The prostitute would seem to be an example of non-organization of touch space. Anyone with the proper fee may touch her – but upon closer examination this is not at all the case. The fee is part of what plays the role of invitation ritual, a business-like shortcut from our usual more complex rites of seduction. Moreover, to her the touch of her customer is very different from that of her lover-pimp. With the "john" or customer the touch is casual. She endures him and rarely experiences orgasm. Not so the relation with the lover-pimp – when he touches her it is "for real." Barbara Streisand's remark in the "Owl and the Pussycat" is perfectly appropriate with the context: "I may be a prostitute, but I'm not promiscuous." Even here the intimacy relations of touch are encircled and constituted through implicit languages of touch.

Thus to change the patterns of invitation and the role of touch in the erotic is also to change the meaning of sexuality. The experimentation which is now occurring in communes and on campuses illustrates this. On an ideological level the "new freedom" is conceived as actualizing a dream of human unity and equality which extends to sexual relations. And surely the boy or girl who lives in this experiential context remains far from the context of the prostitute. In their case mutual invitation within a non-commercial context may occur; a certain type of satisfaction may occur – but if I read what happens correctly there are also antinomies which emerge in the change of pattern.

One factor relates back to the hard to destroy inheritance of a cultural double-standard which still pervades our society. With an abundance of "easy chicks" today's male finds it simple to regress to a crude level of self-satisfaction without a sense of mutuality in relation to his partner. Marx, a century ago, realized that free love in a communal situation could be a merely communal way of treating women as chattel.

A second factor is closely related to the first. If the metaphor of a language is maintained, public or even group language remains a step removed from the subtle complexities of intimate lovers' language. Thus one hears more and more about the breadth of sexual experience and less and less about intimate sensuality – "I've balled lots of chicks, but haven't found many I could love." Non-selective or widely selective sexuality becomes a bit like shaking hands – there are those who grip you strongly with expression of character and others who touch like cold flounders. And while it is impolite to refuse the shaking of hands with anyone, it is more difficult for me to imagine having to share beds with the sexual equivalent of the flounder.

The point I am making is a structural, not an ethical one. The possibilities of multiple dialects within the totality of sexual language is so great as to stagger the imagination. But at the same time within a language or any dialect the most subtle meanings are not generally conferable. What is common remains inverse to what is exceptional. The subtle meanings, hidden significations, much that is unsaid, may easily be conveyed in a dialogue between friends, within small circles, and most clearly between persons who have had the longest and most intimate contacts. This indirect communication *is* possible to a degree even in some wider situations, but not in the same way. To choose one type of language situation as normative is to exclude another, but to choose is unavoidable.

The complexity of our rituals of invitation and the possibilities of organization of our touch relationships is staggering – but they are nonetheless constituted within the varied contexts of pregiven significations. Sexual intimacy unfolds within the "logic" of these contexts. But after all this is said there still seems to remain a sense of primordial privilege to sexual touching. Even apart from the mysteries of desire the full intimacy of touch within sexuality is heightened so that this act stands out among acts of touch.

After the invitation has been issued and accepted, the lovers come together in a play of the full orchestration of touch perception. To overcome the invisible barrier of protection, to invite the other to touch you in this way, to be open to one another, is also to bring into play the full range of touch structures. Two persons, long familiar with one another, who understand the innuendos in the unspoken dialects of touch, retire from the day to end up face to face. Here no elaborate project of seduction nor any complex ritual of invitation is needed because both have long understood the slight movements which say without words.

Disengaged from the worries of the day this retirement into seclusion lessens the ordinary concentration upon the affairs which normally so occupy us that we are likely to miss those fringe phenomena which surround us and give richer texture to experience. The couple, relaxed, enter the bed and bring their flesh together. The disengagement from walking concerns and the surrounding warmth of the situation elicits a field state. The couple embraces through mutual invitation – I surround her and she me with as much of our bodies as possible. In this embrace as we surround each other we maximize the ambiguities of mutuality. We "float" – not unlike the experiences which lower the self-other distinctions – upon each other so that this vagueness of me-other is positively heightened.

Within this context, modulated by the pleasant field state and its me-other ambiguity, further tonalities may emerge. Those sensitive parts of ourselves, within the present and total field, are the focal themes which while not displacing the field state add to it in the symphony of sexuality. But even in these intense moments, even through the focus of the sexual act, the ambiguity of me-other continues. Only the conscious project of the refusal to communicate, to retire into oneself, makes this act alienated. If I inhabit my acts, if I am embodied in the fullness of touch perception, then in this symphonic moment with all of the features of touch fully alive I no longer contact the surface of the other, I live in and with the other.

Now I have opened a way which I never thought I would open. But I have left unsaid more than I have said. Positively, I have pursued a line of thought which sees the fullness of touch expressed in heightened form in sexual touch. The always present and implied intimacy of touch is here explicit and thorough. But this occurrence happens within a complex and highly ordered set of lived spatial configurations. The possibilities of communicative mutuality, implied in touch, are fragile. And although I intend to indicate that the nuanced sexual activity of the lovers who have learned something of the fullness of their "language" of love is a focal possibility, it is also clear that other less nuanced possibilities occur. But if the door to a further exploration of touch within a phenomenology of eros has opened even a little, I shall remain satisfied.

Erling Eng

Psyche in Longing, Mourning, and Anger [*]

> Yet Paradise is shut and the cherub behind us; we must
> take a trip around the world and see whether it is perhaps
> open again somewhere behind.
> – Kleist, "On the Marionette Theatre."

In Book One, Chapter Ten of "Wilhelm Meisters Wanderjahre" the protagonist, stepping away from a telescope through which he has been observing Saturn and its moons, says to the astronomer, "I don't know if I should thank you for having brought this star so unconscionably near. Before when I saw it, it was in relation to all the many other heavenly bodies and to myself. But now it stands out in my imagination out of all proportion, and I don't know if I would like to have the remaining throngs brought up. They will close in on me, make me fearful."

Here Wilhelm describes how drawing the sight of the planet close through magnification is attended with a diminished sense of it as related not only to other heavenly bodies but to the viewer as well. That is, awareness of the other as related to one occurs within a grasp of how the other is situated in a more inclusive, variegated whole. Then, though Goethe does not explicitly say this, it is through comprehension of that whole as virtual, that I am able to realize the relation of the specific other to my own presence, now also more clearly given than naïve sensory perception can ever instruct me.

If other parts of the heavens were to be brought forward in the same merely magnified fashion, the effect would be an overwhelming of consciousness, followed by constriction. The result would be a fragmentation of the heavens as such. The world, though rendered more clear and distinct through the heightened detailing of specific foci of interest, is converted into a throng of disparate particulars, immobilizing the viewer who, his grasp of the whole lost, is no longer able to comprehend the enlarged and distinct details through his presence within a relationship to the whole, but only by way of constructions of theoretical phantasy or mathematical reason.

For many reasons which will become evident this passage from

[*] To Erwin W. Straus on his 8oth birthday.

Goethe provides a useful access to the problematics of eros in a way
that does not isolate it from other realms of experience. While traditional treatments of eros as sacred and/or profane are well known,
something more modest will be attempted here.

Turning back to our introductory text we note that there is an
undercurrent of annoyance, an ever so faint anger about what has
been lost in accepting the spectacular exhibition of the astronomer.
This emerges even more clearly in subsequent paragraphs which, because they are not essential for my argument, I will omit here. If we
read our passage carefully we may also observe that this negative
feeling arises from a sense of having been duped, i.e. in being offered
something marvelous one has been deprived of an originally given
whole which, though less novel, is more sustaining. Adumbrated in
these few lines is a feeling of having been tricked out of such a primordial vision, and there arises a reaction of anger to the loss. Anger in the
form of indignation at the loss overshadows any possible mourning for
what has been foreclosed.

If phenomenology and eros have not been more often discussed, it is
likely that their affinity is so close that it seems difficult to disclose
anything fresh by starting from their relation. A phenomenology of
eros moves rather quickly toward transcendance as eros is discovered
to be as transcendant as Heraclitean fire, of which all creation represents
in its diverse manifestations "measures of fire." But since it should be
evident that I have not settled for speechlessness, nor intend to abandon
reflection for mantic utterance, how to proceed?

We reflect on an exemplary instance of the problem of the truth of
eros (which has always been demonic) that is neither merely human nor
divine, but mediates between them, between what lies within human
power and what we experience as manifestation of a power we cannot
command. The truth of eros assumes the limited truth of mere appearances, no less than the character of all truth as appearance.

The truth of eros has an intimate tie with the problem of the truth
of theatre, even with reference to the old question of whether actors are
artificial or genuine in their expression; thus we may guess that Wilhelm Meister's "theatrical" criticism of star-gazing might nourish
"voyeuristic" meditations on the truth and falsehood of sexuality. This
could be so, even if it were not likely that the root of "obscene" is the
Greek word *scene*, or "stage," the prefix "ob-" having the value of
"against."[1] There is thus an element in the very history of the word

---

[1] Partridge, Eric. *Origins*. New York: Macmillan, 1958. "Obscene," p. 446.

which suggests the presence of that which is in some way "opposed" to manifestation. But, holding the view that such etymological references tend to be implicit representations of perceptions more convincingly represented through consideration of the phenomena themselves, let us return to our first text, which poses in a tacit fashion the question of what a "scene" is, theatrical or otherwise. Because for Wilhelm Meister the view through the telescope was inimical to, and in fact destructive of, the "scene."

What characterizes the "scene"? "'Tis distance lends enchantment to the view" surely applies first of all to the scene. Which is to say that it is first of all a "manifold unity," comprising many particulars, whose unity is virtual. A scene moreover, as is evident theatrically, is shared by a number of viewers. The scene, and this is inseparable from the virtual character of its wholeness, is open to the viewer. This openness is at once its accessibility to the viewer from his position out-side the scene, as well as openness of its depths toward the spectator. The scene is open forward and in depth, in the virtual character of its unity, betokening as well the temporality with which the scene is traversed.

From the scene to the vision of the beloved is not far, as the passage from the *Vita Nuova*, in which Dante describes his second meeting with Beatrice, shows: "Nine years had passed since this first meeting I have described with my glorious lady when, on the actual anniversary, it happened that this miraculous lady appeared to me, clothed in pure white, and walking between two older women. As she passed along the road she turned her eyes to where in fear and trembling I stood. Then out of her inexpressible courtliness, which now receives its reward in heaven, she greeted me so graciously that at that moment I seemed to experience absolute blessedness."

"The hour when she greeted me was in fact nine o'clock; and as it was the first time she had spoken to me, I was so overcome by her sweetness that I left the crowds like a drunkard and went to be alone in my room. There I set myself to reflect on her courtliness."

Located in clock time it is at the same time a commemoration of an earlier appearance, or almost epiphany, so that we grasp it as manifest-ation of a transcendental figure to which Dante longs at once to surrender and to turn away from, as we see from his dream which immediately followed: "As I thought on her, a gentle sleep fell upon me in which I saw a remarkable vision. I thought a flame colored cloud appeared in my room, and in it I saw the figure of a lord, terrifying to anyone who should look at him. The delight he seemed to take in himself was quite strange. He said many things to me, of which I understood little, amongst them being: 'I am your lord.' "

But in contrast to the scene of the day preceding, the dream now becomes fleshly vivid and heated in its imagery: "In his arms I saw a woman sleeping, naked apart from a blood-coloured cloth lightly wrapped around her. As I looked at her closely, I recognized the lady of the salutation, who had greeted me on the day before. In one of his hands he held something that was all on fire, and he said, 'Behold your heart.' When he had stayed there some time, he woke the sleeping woman and forced her to eat the burning thing in his hand; and she ate it, with great misgivings. In a short time his joyfulness changed to the bitterest weeping; and so, lamenting, he clasped the lady in his arms and turned away with her up toward heaven. At this, my anguish was so great that my light sleep could not withstand it, and I awoke."[2]

The contrast of Dante's dream with his waking vision is similar to the contrast between Wilhelm Meister's telescopic view and that of his image of the nocturnal heavens granted him by the naked eye. Instead of a scene that is shared, it becomes one that is private; rather than opening up it becomes constrained; rather than a vision of unity it tends toward the fragmentation, literally, of cannibalism. All these negations of the characters of a scene justify its being termed "obscene."

The obscene is to be understood from the scene as norm, as, in effect, a withering of the scene in its fullness, from virtual unity to breakdown into concreteness, from openness and depth to constriction and raised figures, from shared experience to one increasingly autistic. Such experience of the obscene is more common than is realized. What is ordinarily termed "obscene," however varied it may be, in accord with its autistic tendency, is merely that on which it is possible to secure a kind of rough consensus. But what can be agreed on as being obscene becomes, to the extent that it is defined by consensus, already counter-obscene. Pornography is an instance of this. Thus the attraction of obscenity becomes public and, like a grisly accident that attracts crowds, by its public character already negates the reality of privately imagined fears. It is from this that the apotropaic possibilities of the obscene gesture or word derive. The ceremonial public performance of obscenity can represent a way of countering the most per-

[2] Dante. *The New Life*. Penguin Books, 1964.

sonal fears, as in those end of the yearly observances in the Saturnalia, Carneval, Fasching, Mardi Gras, not to mention the ceremonial obscenity enjoined for occasions of mourning and separation among African peoples.[3]

The obscene is not merely the obverse of the scene as already described. As we see from Dante's words it is also a modified reproduction of the past of a particular scene, whether the past be a verifiable memory or a mythic form. Such a mythic form has the character of a timeless whole, into which discrepancies among seeing, remembering and imagining have not yet entered. For Dante it was his first glimpse of Beatrice: "She was just eight when she first appeared to me, and I was almost nine. She was dressed in a noble color of subdued and plain crimson, girdled and adorned as suited her early years. At that moment, I am speaking in all truth, the spirit of life which lives in the most secret room of the heart began to vibrate so fiercely that its effect was dreadfully apparent in the least of my pulses; and, trembling, it said these words: 'Behold a God who is stronger than I, who in his coming will govern me.'" But at one and the same time (as Dante proceeds to tell us) he is torn apart by dissension among his emotions, intellect, sense perception, and instinctuality. For just what binds is discovered as severing. Psyche, enamoured of Amor, suffering separation and disunity of self, strives to become free of the bitterness of her delight. Dante's effort to penetrate the meaning of his plight takes the form of absorption in the image of the beloved to the point where it discloses the figure of Eros himself as diabolical; then enigmatically, as we see in Dante's dream, the obscene supervenes. What can this mean? We turn to the tale of Amor and Psyche in *The Golden Ass* for help.

It is the story of Amor and Psyche which reminds us that we have failed to give sufficient consideration to the scene as a visual experience. The scene is a view, even a "vision." The central event of Apuleius' tale is Psyche's transgression of Amor's injunction that she not try to look at him while he visited her nightly: "He warned her with terrifying insistence that her sisters were evil-minded women and would try to make her discover what he looked like. If she listened to them, her sacrilegious curiosity would mean the end of all her present happiness and she would never lie in his arms again." But her sisters' envy and, even more deeply, their fear brought her to the fatal step: "Psyche's curiosity could be satisfied only by a close examination of

[3] Evans-Pritchard, E. E. "Some collective expressions of obscenity in Africa," *Royal Anthro. Inst. of Gt. Brit. & Irel., Journal of*, 59, 1929, 311-331.

her husband's sacred weapons. She pulled an arrow out of the quiver
and touched the point with the tip of her thumb to try its sharpness;
but her hand was trembling and she pressed too hard. The skin was
pierced and out came a drop or two of blood. So Psyche accidentally
fell in love with Love. Burning with greater passion for Amor than
even before, she flung herself panting upon him, desperate with desire,
and smothered him with kisses; her one fear now being that he would
wake too soon."[4] When a drop of scalding oil fell from her lamp on the
sleeper's shoulder, he awoke and left her, as he had warned. Since it
was the "curiosity" of Psyche which had brought about her downfall,
we are not surprised to learn that the final test of Venus she must undergo for her to be rejoined with Amor is to bring back a little box of
divine beauty from Persephone in the Underworld without ever looking
into it.

The Apuleian "curiosity" which proved the undoing of Psyche is
not far removed from Goethe's indictment of telescopic astronomy as a
kind of celestial voyeurism. In both cases the result is a sense of loss,
mourning for a now lost fulfillment, and indignation about what has
happened. The last is nicely expressed in Apuleius' fable by a gull who
complains to Venus that as a result of her son's unfortunate affair
"Pleasure, Grace and Wit have disappeared from the earth and everything there has become ugly, dull and slovenly. Nobody bothers any
longer about his wife, his friends or his children; and the whole system
of human love is in such complete disorder that it is now considered
disgusting for anyone to show even natural affection." In the dream of
Dante, by contrast, the sense of loss and mourning is more evident than
the anger, though the heat of anger may be felt in the force applied to
the sleeping woman to compel her to eat the burning heart. The curiosity, i.e., eros as preponderantly visual, results in a loss of the situation of
the whole, with loss, anger and unassuagable yearning, a kind of incen-
diary revenge not without its own magic, as we see in the *Vita Nuova*.

It is a possibility of eros that it may be confined to fascination with
a particular person or even situation. And drawn into this one, there is
an isolation of the beloved from the rest of the world, and even an
isolation, a kind of inner separation, of the lover from himself. Thus
eros has been considered conducive to a kind of madness its disloca-
tion of world and self has brought about. Acquaintance with eros soon
leads to the discovery that its fullest meaning is somehow precluded by
its actual identification as such. Then follow the insensate attempts to

[4] Apuleius. *The Golden Ass*. Penguin Books, 1950.

recover the solution by gazing upon the enigmatic image in which eros in all its nakedness seems incarnate. Yet inexplicably such focus only intensifies the anguish of a loss in which thought and feelings commingle. It has not yet been grasped that the image of the beloved is more than an occasion for meeting one's own desire. As long as this is not realized, that image is largely a mirror for oneself, as Psyche saw only her own desire and fears in the hidden figure of her unknown husband. Thus the desperate attempts to grasp the meaning of the erotic image as mirror must result in a progressive dismemberment of the image of the other as well as of oneself, culminating in the final discovery of a rupture between Psyche and the world.

The preoccupation with the obscene, in one or another way, is an attempt to enlist anger in the service of love. To the degree that desire is given primacy over presence to one another, as if desire as known in consciousness were constitutive of it rather than the other way around, to that extent love is corrosive alike of beloved and lover. In these circumstances isolation of the object as object of desire from all the world around tends toward its decomposition as well as toward my own decomposition in my dependency on that object. Psyche, suffering an enigmatic affliction of poverty amid apparent plenty, has not yet understood Amor as disembodiment and re-embodiment, as commemoration. Without that realization the body of the beloved becomes – as in the fetishistic doll of the Surrealist Hans Bellmer – a machine for the gratifications of an incorporeal phantasy, in that vision which is sheerly, in the words of Apuleius, "the Desire of Desire."

II

The obscene, in the foregoing perspective, appears as a kind of "catastrophic reaction" of Psyche in the presence of Amor, in which fullness of presence is crippled by the magical, self-reflective use of images. This is, in the first instance, a private vision, something like a dream; thus, when expressed it is either absurd or anxiety provoking for others. Ordinarily we consider the obscene as an expression which violates the norms of what is to be shown bodily or referred to in public, and if so shown, to have a violent connotation. At the same time the obscene is evidence of failure to realize the peculiar limitations of the "scene" in love. Thus a consideration of the obscene in its various forms is helpful for understanding "errors" in the development of an awareness of eros as commemoration, of eros as preservation and

origination alike. It is an implicit interpretation of eros as solely originative or solely preservative which produces those aporias in experience
which are part of the obscene.

To consider the implications of this view, the obscene will be briefly
considered under four different heads: anger, social protest, regeneration, and madness.

An angry person may, in cursing, refer to intimate parts of the body,
even point to or use parts of the body ordinarily not so referred to or
used. The true curse is accompanied with something like a magical
intent to harm another through the force of one's utterance or gesture.
In this the sense of liberation which is part of the feeling of the obscene, of becoming freed from all restraints, becomes evident. Thus
"obscene language."

Social protest as obscenity is however additionally an implicit,
iconic denial of the legitimacy of the prevailing cultural norm or social order. By exemplifying what is tacitly forbidden in the prevailing
norm or order, it is a silent indictment of its rightfulness, when, for
example, the national flag is reduced to a material from which a phalli-
form figure is shaped. Characteristic of some recent "Pop art," as well
as public exhibitions of nudity and defecation, it is often accompanied
with a strong display of moral indignation. Historical antecedents of
this kind of obscenity are to be found in satire and caricature.

Obscenity of regenerative intent was a feature of celebrations which
brought a year or other period of public time to an end and were the
start of a new period. Now largely disappeared, they included the
Saturnalia, Carneval, Fasching, Mardi Gras, and similar festivities.
During these times the usual order of things was inverted: the socially
high exchanged places with their social inferiors, and men and women
exchanged parts and intermingled freely. With the ending of the year
as a constituted period of time the constituted social and cultural
world reverted with it to an undifferentiated condition through sanc-
tioning a *mundus inversus* which, relative to the constituted order, had
the value of the obscene. But this obscene was of a sacred, i.e., regenerative character. Its quality was a drunken one, and was assisted with
intoxicants, whether drugs or alcohol.

Madness is capable of disclosing the obscene in its purest form. I
have known a hospital patient who felt that she had been turned inside
out like a glove which had been reversed, so that she was exposed to
the world with intense feelings of shame, her viscera, mucous memb-
ranes and bodily fluids all on the outside. This is the most extreme

form imaginable of that helplessness and exposure to the world which is faintly suggested by the worlds of Goethe in our earlier quotation: ". . . they will close in on me, make me fearful." Here this violation by the world of the person has passed through her body as a kind of vanishing point, so that she now exists solely in the eyes of others, all but empty within, and full only without, in the gaze of the others.

All these four forms of the obscene hold a kind of magical destructiveness; even the psychotic woman I have mentioned had some sense of overwhelming others with the feeling of her strange condition. This power of destructive feeling is intensely fascinating and has an erotic quality of its own. It is the power of a failed and envious eros, as we see in the embittered, scheming sisters of Psyche.

III

The reflective reader of the passage from the "Wanderjahre" may have noted that the question of "scene" is involved with Wilhelm's "theatrical" reflections on astronomy with the telescope. We are not far from the world of theatre and the problems of the actor. In the obscene we create or are invited to create a spectacle for ourselves, a kind of theatre of erotic consciousness in which we are at once players and audience – not to mention being the writer of the scenario. We are able to do this because of the "duplicity" of our experience as bodily. On the one hand I "am" my body, on the other I "have" my body. In either case there is a kind of body as an other; in the first case my bodily complement is in or is part of the world, in the second instance I refer to the body as "my own." Although these two modes of awareness usually complement one another smoothly, it is also possible to experience one of them as overlapping and obscuring the other. Now one is set against the other, in conflict. If I use the image of the other as a kind of mirror in which my own desire is reflected, then the other person can also use his image to affect me in accord with his or her wishes. Or if I do the same to another person who uses particular images of others to receive back his own desire magically, my action is in a very real sense a violent one, and if it is sexual, then we very readily characterize it as obscene.

Here the obscene is a kind of magical violence which penetrates considerations which would interfere with a less "manneristic" accomplishment of relationship. We find evidences of this implicit "rage to love" in those mutilations which, formally and thematically, are to be found in obscene expressions. Here we see the obscene as an attempt to convert the anger of failure into a magical erotic resource. This resource is used to establish an illusion of fundamental communication simply through the "bodies we have." Such a formulation makes evident the irrelevance of the person in such a situation. The attempt to secure complete immediacy to one another simply in terms of the "bodies we are" is no less doomed to failure, since we are always there in some measure as "bodily otherwise" for one another. This is but the equal and opposite alternative of erotic utopianism, of which there is not a little in Dante's account in the *Vita Nuova*.

The situation of eros carries within it an import of myself as "body I am" to the beloved, suffusing the beloved (who has appeared to me in the mode of possible possession) with the meaning of "body I am," or "I." At one and the same time I discover myself as having already appeared to the beloved, beyond any possibility of being a mere possession, as "body I am," or "I." This is the possibility of mutual fulfillment, one of whose forms of default is that of the obscene, in which the distinction between the modes of consciousness of "body I am" and "body I have" are partially experienced as overlapping, and are at once suffered and magically used. To the extent that a relationship includes such an emphasis it is also a struggle between partisans rather than partners, in which each one's fulfillment is exclusive, each competing with the other for what he believes rightfully his own, of which he has been wrongfully deprived, each failing to accomplish that meaning which can come to neither alone because it belongs to both. A situation that was once open has been converted into an issueless strife in which each sees the other as possessing what he himself or she herself lacks to attain that completion which both – together and yet apart – vainly attempt to seek. An unrelenting envy of possession thwarts the possibility of realizing gratitude in simply being present to and for one another. The envious sisters of Psyche, who excited her erotic curiosity, are also Psyche's intolerance of the difference between herself and her lover, difference which remains the very condition of eros as well as of time, in which present fulfillment commemorates past ones and promises a future.

IV

It has already been observed that the obscene is closely tied to the visual sense: it is the glimpse of her sleeping spouse which brings about

Psyche's loss. The significance of vision is that it emphasizes the merely corporeal distinctness of the other, tending to be supportive of an anticipation of fulfillment in terms of mutual bodily possession, a notion which is absurd, given the unilateral character of the relation of possession. Love, wary of the separative mode of vision and of light, seeks privacy, twilight, and even darkness, all in favor of those senses which, like touch, smell, and pressure, are more generous to the admission of the other. Whenever vision is stressed in love there is a tendency for it to gravitate in the direction of the obscene. In suggesting the eventual possibility of a completely visual equivalent of eros, an "optical orgasm," the never ending effort to reveal ever more moves by degrees into the penumbra of the obscene. This is a virtual tropism of photography and cinema alike, to the extent that their reliance on a naive positivistic notion of vision provides a basis for their attempt to "disclose" the truth of love in visible terms alone. From the peripety of such convulsive efforts at reconnaissance it may eventually return with an enhanced awareness of the importance of what cannot be simply shown for realizing presence to one another. From the cryptic suffering and violence of the obscene there is always a chance of returning with a clearer recognition of the eventual necessity of a language of gentleness and consideration. But this must call on the strength to abandon magical pretentions which have been seen through as specious.

These are the magical obscene expressions earlier considered: anger, protest, regeneration, and madness. When employed angrily the intent of the obscene is to injure the other in that part of his body to which the curse or gesture alludes, even while apparently oblivious that its efficacy depends on a oneness of bodily being. When used in protest the intent of the obscene is to jab, shock, provoke the body politic awake, as a kind of alarm, an intention often indistinguishable from the merely angry wish to hurt. The obscene of the Carneval and other regenerative observances aims at a new beginning of the world following reduction of the old order. Finally as madness the obscene is revealed as a magic attempt to materialize the other in one's own image, to congeal the ambiguous sense of the living environs, to control it in and through the dread of one's own present. Tenuous as the histrionic moment may seen in the insensate rage and grimaces of some psychotic patients, it is yet evident that they include the intent to consternate others by using dread to paralyze their living openness, subdue their disconcerting metamorphoses. Minimally, it is just this intent which grounds the apotropaic use of the obscene gesture or figure.

In preliterate societies it is often the shaman, in one of his many forms, who has the task of confronting derangement, converting it into sickness by taking the responsibility for reducing it upon himself. That he must confront the obscene, solve the riddle of its magical allure, is suggested by the part played by the ritual mastery of the obscene in the initiation of future shamans.[5] In the course of becoming a shaman the future healer may also experience a petrifaction, followed by dismemberment or shattering, of his own body as the prey of the demons of illness, an event which enlightens him as to the service others will subsequently expect of him, namely to dispose of their sickness after it has, in one or another sense, been shifted onto him. He faces the obscene on the threshold of the passage from health to illness, and a second time in the movement from illness to health. Although its apparition holds the promise of change, this possibility is, first of all, a baleful one. Its most immediate meaning is an intimation of death in doubt, weakening, or failure of living trust, a doubt at which Wilhelm Meister's remarks on the possibility of voyeurism in astronomy hint, Dante's distant worship of Beatrice, or the failure of trust in Psyche's subservience to the dread of her sisters. In longing Psyche mourns the loss of a love whose completeness is revealed in retrospect, casts about in anger and tries to place her blame for having failed to obey Amor, everywhere suffering the obscene. Little does she realize that the obscene which she everywhere encounters is the mask of love become concretized, its expression the evidence of her own malaise; it is, in André Breton's phrase, "convulsive beauty," the image of Narcissus twisted in its own mourning and anger. That it may be also held up to others as a mirror to expose to them their own narcissistically faulted ways is a possible use for it, however problematical for the user such use continues to remain. It remains problematical because of the essentially magical character of the obscene. By magical I mean the activation of the bodily mimetic imagination. At this level the four uses of the obscene I have distinguished are not quite so clear. Indeed, any one form also implies the other three possible meanings in the perturbed relation of Psyche with Amor, whether the relation be one between individuals or is a relation of the group or collective. Groups, too, are capable of longing, mourning, anger, and fascination with the obscene.

[5] Makarius, Laura. "Ritual clowns and symbolical behavior," *Diogenes*, 69, 1970, pp . 47-51.

The "problematical" character of the social or critical use of the obscene derives from the way in which its usefulness is based on a tacit identity of those who use it to show default and those whose default is to be pointed up. Thus a kind of collusion is present, which is at the same time apparently denied by the vehemence of the would-be exposing gesture. That those who hold the mirror up do so without acknowledging co-involvement in the default is at the same time a denial of the more basic tie on which the efficacy of the warning depends, i.e., if it is to be heard and accepted as a warning. Gandhi knew this, and it was the salt of his Satyagraha. Denial of this deeper unity of existence is apparent in the narcissism and violent instinctuality of the self-righteously obscene. At the same time those who would deny the reality of the obscene in living are also serving another narcissism which would obscure any allusions to the struggle to maintain existence against the possibility of its failure. The deliberately obscene display retains a destructive intent, quite in contrast to the obscene as inadvertently encountered in the everyday world.

There remains the possibility of a deliberate evocation of the obscene which is meaningful. This would be the equivalent of an exorcism. We have referred to this in the examples of the regeneratively obscene, especially in connection with healing. This has usually been considered "catharsis." Without going into a detailed consideration here, one point needs to be made for our discussion. A deliberate encounter with the obscene may be fruitful on condition it remains within a larger frame whose legitimacy remains untouched by the indictment of the obscene. Whether obscenity of year's end observances, religious mysteries, or healing experiences, it would seem productive only in the context of conditions which serve to keep us mindful of the provisional character of its metamorphic disclosures. The contrary is the case when the encounter with the obscene is sought by those who lack either ability or the conditions or both for the indispensible reflection that can disclose its significance as the distorted commemoration of an obscured catastrophe in loving. Then either a retreat with their fragmentary revelation, or terrorization of others in the way they were terrified, must follow. This kind of self-perpetuated impasse goes far in accounting for the addictive character of the obscene. And it may be usefully related to other forms of addiction, all of which seem so intimately involved with the disturbances of Psyche and Amor.

V

It is above all the image, sudden, violent and arresting, which is characteristic of the obscene. Consideration of this image may assist our understanding of Psyche in longing, mourning and anger.

The erotic image is at once the place in which I meet the evidence of my own desire as well as the other appearing within consciousness to me in that desire. The image is the place in which my desire materializes so that it now becomes the place from which my own desire appears to me, as we see in dreams and daydreams alike. At the same time the image remains open for the other to appear in and through it. But to the degree that I am fascinated by the materialization of my desire in the image of the other, the image serves as a barrier to the further entry of the other. Eventual further emergence of the other as other, through the barrier set up by my own desire materialized like a mirror and reflecting my image, must necessarily disrupt my pleasure and my reflection in the form of the other's image. The intrusion of the other as other, incompatible with my own egoform shaping of the other, must disappoint and eventually antagonize me. Should I remain attached to these memories of the beloved ego and its reflex, longing replaces presence, mingled with mourning and anger. This possibility of reflecting my desire so that it becomes a self-originated possession, through using the image of the other as a mirror, makes it possible for Psyche to avoid facing Amor, excluding Amor by remaining an observer. Amor cannot meet Psyche through the image because of the barrier of Psyche's desire, which now modifies everything appearing there in terms of Psyche's image of her desiring self. This is the condition of that "Desire of Desire" which Apuleius describes as the affliction of Psyche, realized after she had looked on the naked Amor. Which is to say, that light cannot take the place of vision, and that when this is attempted, we but receive back our own image.

It is just this partial opacity of the erotic image as invested with our own desire, as a mirror in which we may meet our own desire, obscuring the other as such, which endows it with the character of an abstract schematism. This abstractness is not always recognized as such; indeed, the obscene gives the impression of extreme concreteness. But we can now understand this concreteness as an effect of meeting with one's own desire, while its abstractness results from its limited accommodation to the reality of the other within whose image desire has materialized. This abstractness accounts for the monotony of pornography, a special case of the obscene, and for the absence in it of any reference to faces. The guiding scheme is an image of the other in which one's own desire remains reflected and apart, intrinsically opposed to a progressive manifestation of the other person.

Since a materialization of one's own desire in the image of the other is opposed to the presence of the other as such, there remains implicit in such a use of the image of the other – a turning of the image of the other as apparition against the appearing of the partner – a mourning for an inexplicable loss, a longing, and an objection to the other as a kind of obstacle. Thus the image of the obscene conceals an anger against the other whose image it makes use of, and it is such anger which accounts for the "unexpected" explosive consequences of introducing the obscene. The characteristic irridescence of the obscene arises from the way in which desire is deformed by an unaccountable anger, giving the obscene expression its often self-justifying, self-righteous intensity.

A desire turned back on itself remains committed to an object of desire which increasingly turns out to be an enigmatic mode of one's self, a form of the "double" as it appears in the narcissistic pursuit of the self in the forms of its reflection. In this perspective the psychological is concerned with the crises of narcissism as these occur in the ordinary course of living. And the psychological discovery par excellence is that of the recognition within such narcissism of implicit anger toward the beloved, whose image has served as the place of the *malentendu* and materialization of desire which has become one's addiction to the "Desire of Desire."

Such an erotic anger exercises an imperious attraction, since it betokens that missing other who has been sacrificed to Psyche's reflection of her own desire and its fears. Is it surprising that it can be used in one last attempt to end that failure of love pictured in the obscene by a destructive attack of Psyche on a world that seems to have destined it to failure? We are told in Genesis that Cain was also the founder of civilization, its arts and crafts. We might account for this by imagining Cain returning from the narcissistic anger in which he slew his brother, revulsed by the disclosure of his delusion, to found an order in which narcissism is allowed to assume other, less violent forms – at least immediately.

The intimacy with which anger commingles with desire in the obscene is provocative of further reflection. It is remarkable that the unknown bridegroom of Psyche is initially announced as one "Who flies through aether and with fire and sword/Tires and debilitates all things that are" and concerning whom Psyche asks, as she is led to her

nuptials: "Why should I shrink from him, even if he has been born for
the destruction of the whole world?" The meaning of Amor hides a
mysterious dread. It seems that it is Psyche's exceptional beauty that
has condemned her to separation from her royal parents. But the separation is in the service of Amor, an intimation attended with dread
for Psyche. What the child loves it rejoices to rediscover. That this joy
of rediscovery should mean anything else but an everlasting profusion
of riches, that it means loss as well as abundance is as yet unguessed.
Now, unless love of the new is lived together with loss of the old, and
there is learning of commemoration, Psyche is condemned to wander in
longing, mourning and anger. Given our limitations, love compels
us to accept the loss of earlier beloveds, if love itself is not to die. Amor
is conflagrative as well as nurturant. Denying this, Psyche becomes the
prey of anger, which in the tale of Apuleius is the anger of Venus as it is
twice turned against Psyche.

"Infantile" eros, oblivious to the time in which it takes wing, turns
in alarm to preserve what it has inexplicably seemed about to become
faithless to. Refusing to accept what is happening, its own possessiveness
is enhanced, and as its struggle to maintain its treasure unchanged
proves unavailing, its anger mounts. Failure to penetrate the initially
fearful aspect of its presence, while emphasizing this aspect as preservative of the past, is to render it a possession. The fulfillment of Amor
and Psyche exacts relinquishment of the infantile "I," that possessive
infantile "I," strengthened in its illusions by all the feints and incon-
stancies of Amor endured during the most magical years of being human.

If we now look back to the opening passages from Goethe, Dante,
and Apuleius, we may be surprised to realize that they all involve the
presence of the night as a time of disclosure. Both time as the new day
and Amor appear in the night. Within such a context what we have
called the obscene seems an attempt to render in terms of day what can
only appear within that night in which time itself and the new day already emerge. The obscene is a "minotauric" image, a negative reciprocal of the possibility of realizing a perennially actual eros.

Now there emerges another sense of Kleist's observation that with
"Paradise shut and the cherub behind us, we must take a trip around
the world and see whether it is perhaps open again somewhere behind."
It is perhaps open again "somewhere behind" in the grace as well as in
the act of memory, which is both a realization of the depths of present
love and a recognition of it as more than a repetition, indeed as a commemoration of past loves.

HENRI VAN LIER

Signs and Symbol in the Sexual Act

The human being lives in the midst of signs. One can even define the human being as the originating principle of signs, and with structuralism describe human consciousness as the void, the empty box through which signs change and eventually reorganize into new signs.

Characteristic for the sign is that it emerges, rises from a ground; at the same time it takes on integrative properties. Thus it presents the essential features of form (*Gestalt*). But beyond the fact that it is more conventional than the form, it joins to the perceptual datum a concept; this is made up of a signifier (for example, a word) and a signified (the concept designated by this word). On the other hand it refers to its environment in a definite manner: a sign always refers to another sign, says Peirce; in the world of signs, says Saussure, there are differences only, so that a sign signifies only within a system of signs; it gives rise to a discourse in being enunciated, or at least to a course in being effected. Language is assuredly the most elaborate network of signs; but it is necessary to bring signal systems together, including all objects, artificial and natural, whether seen through languages and scripts, or when presenting themselves as scripts. One sees how the world of signs, rich and ambiguous as they are, places the human being in the presence of structures at once precise, intelligible, and effective. But these structures are relative, i.e. localized and successive, spatio-temporal. And they remain before the subject, like objects (*ob-jecta*). In short the sign presupposes, in every way, abstraction. Making at once for lucidity and action, it furnishes the stuff of empirical and scientific knowledge, as well as of practical life and social organization. Which is to say that it has attained its purity only in the West. But even where its status remains uncertain, as in those cultures not deriving from technical Greece, it evidences, relative to nature, a discontinuity and detachment sufficient to demonstrate humanness.

Yet it seems that the human being is not content with this realm of

signs. With non-Western man one discovers even in everyday life, and
with Western man in certain exceptional activities, experiences of the
absolute, let us say total and immediate, abolishing space and time,
the opposition of the subject and the object. These experiences are
three in number: the sexual act, artistic rapture, and the mystic illu-
mination. In each one the nature of signs is profoundly altered. The
body of the beloved and of the lover (or their duality); the major work
of art, in its different forms: poetry, painting, architecture, music, the
dance; the Self to which the "interior castles" lead, none provide any
relief over against what surrounds, neither articulation of the com-
ponents, nor the rigorous placement in a system, neither the possibility
of strict ways of designation nor of performance, neither of objectivity
before the subject, nor even distinction of signifier from signified. On
the contrary, these phenomena, each in its own way, abolish or encompass their environment, and are thereby infinite; here the parts
are already the whole, and here the whole resonates as such in the
parts. Thus they vanquish space and time; the signifier comprehends
in itself the signified; the opposition of subject and object is surmounted; the active and the passive merge spontaneously. Consequently, even though there is a meaning here, one cannot speak of
signs, nor of signification, and it would have been easy to reserve for
this purpose the term of "symbol," in opposing it to the "sign." But
"symbol," "symbolic," "symbolism" often designate signs and the
capacity in general of signifying. One ought then, to avoid confusion,
each time indicate that the experiences of the absolute involve symbols
which are "plenary," or "radical." We have nevertheless decided, in
this study, to term *symbol*, merely those manifestations of meaning
which exceed – in both directions – the world of signs.

The word "symbol" has the advantage of marking the fact that the
theme of the experiences in question is not entities but relations –
which still does not go beyond the signs – but where the relations in
question are not posterior to their terms, where it would even be inadequate to say that their terms take place within. On the contrary,
like the Greek *symbolon* (an object divided into two parts between
two separate persons which enable them to be identified, according to
Bailly), the symbol is a relation anterior to its terms, engendering them,
a unity whose partition is simply a fructification, an internal polariza-
tion. This conjunction, prior to what are conjoined, governs in two
ways what we are discussing. On the one hand, it defines the relationships between the parts and the whole in the object: the way in which

in the beloved body, in the work of art, in the mystic Self, the parts do not remain parts (even of a Gestalt) but are internal resonances, and in each instance integral ones, of the whole. On the other hand, this conjunction rules the contact between the object of the experience and its subject, it brings it about that there is no longer a subject contraposed to an object (*Gegen-stand, voor-werp, pre-met*), but precisely a couple, where the lover and the beloved, the work and its beholder, the Self and the "I," find themselves literally produced by a unity anterior to their distinction, a unity of which they are the dialecticization, lacking which it would stagnate in identity. These two aspects harmonize: it is because the parts and the whole are completely resonant in the object that there is a coupling of subject and object, and reciprocally. In brief, the symbol is absolutely concrete.

This shows that it is not *some thing* which is total and immediate, and that here there are not so much symbols as symbolizations. In particular concords (or discords) of subjective and objective conditions, in certain of their dialectical moments, it is evident that perception, instead of remaining perspectival and successive, as in ordinary life, anticipates, in each profile of the object, its whole. The imagination no longer consumes images at once delimited and successive, materials of memory and need, but phantasies, matters of reminiscence and desire, or more precisely fantasy, by which we designate (in a usage a bit different from that of psychoanalysis) the primordial and individualizing union of a conscious body with the world. In short, the accomplishment is no longer a work, which produces or constructs; since it displays the unity instead of effecting it, it proves just as passive as it is active; or better, as spontaneous it passes beyond the opposition of activity and passivity by the same movement in which it surpasses that of the subject and object. Now, perception, imagination, realization, became immediate and total, not only enable one another's possibility, as in everyday life, but as vital aspects of one another, as aspects of a body given up to its pure presence at the same time as given up to the pure presence of the universe as whose focus it experiences itself.

One must not divide the sphere of symbols from that of signs. The symbol is a negation of determination; but it has been observed, ever since neo-Platonism, that a negation of this kind has no meaning save by the determination which it negates; it is the richer as the negated determinations become more numerous and subtle. Thus the sexual symbol, artistic and mythic, requires a world of signs, a world of everyday life or of science if its negation is not to be an empty one. Just as, reciprocally, the world of everyday life and of signs has need – existentially if not formally – of the underlying presence of symbols to preserve the living tie without which relations become lost in their unending reflections.

One could define coitus as the symbolization obtained through orgasm in completing a caress that crosses the bodies of two subjects according to the modalities of masculine and feminine.

Physiologically, the orgasm involves a synchronization of neurons rising sharply toward a climax before breaking down in energy discharge. Psychically, it unites in extreme fashion intensity with expansion. Discriminating nothing, it transmits no information, it neither receives nor sends any sign; it is at once perception, imagination, motor activity, none of which are informational, and thus indefinite, infinite, like the rest of the genital sensation which prepares it. Finding its point of application at the center of the organism, it brings it together again as a whole, securing the fusion of its parts, and at the same time their diffusion. Experienced from the abdominal region, and from the extremity of this region, it leads the living body back to the level of its continuity with nature. It is even an experience of fusion, from the fact that it takes its felt start in the erectile tissues, leading to an effusion, projecting the subject outside of himself in a manner referred to as his ecstasy, so much so that the play of sensation and desire here becomes the sensation of desire. Its passivity, or if one prefers, its spontaneity, is such that it is able to pass for an automatism. Its predispositions for symbolic experiencing are evident.

Nevertheless, if the orgasm were nothing but that, it would be confusion, loss of consciousness; its renditions of totality and immediacy would be empty. But it is rhythmic, i.e., return at the same time as departure, coming and going. This can be seen in the failures of the male: overly abandoned, he loses himself in premature ejaculation; overly restrained, he hesitates in delayed ejaculation. And there are the comparable difficulties for the female. As Ferenczi has well noted, orgastic success is a subtle equilibrium of abandon and recovery, of transcendance and immanence, of ecstasy and consciousness, of indetermination and determination. Projection to the extreme limit of self, but remaining within the self. Thrusting and self-control in turn, alternating so rapidly as to form but one consciousness, one *Erlebnis*.

But, because of such seizure the orgasm, too brief, would remain incapable of the symbolization, if it were not joined to the caress, which more tranquil, prepares the orgasm and culminates in it. The caress also has two poles. It is a procedure of fusion, of a joining of parts, of a reciprocal generation of parts, or, more precisely, their derivation from their antecedent unity; it thrives on vertigo from the moment it becomes sexual (the simple familiar caress does not concern us here); it does not distinguish subject and object, but invites them to emerge from their duality; culminating at the genital focus, it leads the organism to melt and radiate from its very center; it never explores, manipulates, nor discriminates; it makes use of perfumes, of dusk, of prattle, the better to flow. But at the same time, while dissolving determination, and in preventing its perception as such, it yet introduces it, because, however its ways melt and dissolve, they are still, however slightly, its ways. Inasmuch as it is attentive to the least detail, even if it is always the whole which is perceived in this detail, the caresser summing up the caressed in the lobe of an ear. Moreover, the caress develops the hint of seduction, even if the subject abandons himself. Thus the caress, determining indetermination, has the same antinomic structure as the orgasm in its coming and going. Likewise it connects more of the finite, the partial, the successive, to the infinite, the total, the eternal – but from within, without compromising them, so that they are not lost in indifferentiation.

Thus it is necessary to see that the caress and the orgasm are two moments, two aspects, of one and the same experience: the caress is the orgasm begun, guided, differentiated; the orgasm is the caress accomplished, fulgurant, ravishing. Thus both ground the ambiguity of the symbolization, each within itself, but in their relations as well. For the articulation of immediate and mediate, of conscious and unconscious, indispensable for the symbol, is realized by the caress in favoring each first term, and by the orgasm in favoring the second.

Nevertheless, however close and fluid the object may be, the subject, if it did not receive any response, would remain in front of it, and the exteriority, the opposition, the activity proper to the world of signs would not be truly surmounted. In truth the sexual object is itself feelingful, and feelingful to the same caressing and orgastic rhythm as the subject. So that the sensation of the one, instead of terminating in something strange, drawing back or dissipating, returns to itself via the sensation of the other, achieving closure. Thus the partners, face-to-face in a circle of feeling together, enclose a world, including in it the world and the consciousness of world. Nor do they need to be active, each one taking birth from the other.

However, reciprocal touching must not simply join like to like. For then the circle of feeling, returned from the identical to the identical, would revolve without reference, or, giving each one a response merely the equivalent to his question, would not enable egress from the self. The simple desire of the other's desire, if the two desires are not qualified, ends in a tourniquet and does not overcome solipsism. Only the coital couple does not confront the same with the same but refers it to what is nearest itself. In effect, if one reflects on the minimal relation where affective perception, imagination, and sensible performance – which alone are in accord with the concrete design of coitus – are able to make a unity manifold without distending it, one finds the relation of tenon and mortise; every other way of dividing the initial whole yields a lesser intimacy in their act and in their result. Now coitus not only actualizes this relation but makes it the focus of its attention. It is the relation of tenon and mortise now become sensible and the primitive theme of desire. The joined partners, and, moreover, only face to face, do not have a relation between themselves which binds them in a unity; rather they diversify, at the least possible remove from the origin, the anterior unity of which they are the parting. The genital feeling is experienced as included by the mediation of the includer, as including by the mediation of the included, and in such originary fashion that it is cosmogonically lived.

Thus coitus does not favor the prevalence of the masculine over the feminine, nor the inverse; and it is very much to the point to ask if there is but one libido, masculine, as Freud thinks, or whether there is a second, feminine, in the view of Melanie Klein. On both sides there is but one desire, one libido, that of Conjunction. The coitus is the Conjunction, a unity anterior to its terms diversifying itself, polarizing itself in the roles of fleshly tenon and mortise, according to the anatomical and physiological possibilities of each.

This anteriority of copulation (at least phantasied) to its partners is so genuine that there are not, properly speaking, sexual organs, i.e., apparatuses which precede the sexual act, as the hand precedes grasping, the mouth chewing. The penis does not become an organ save in its erection, the vaginal orifice in its opening, as vaginismus attests. It is not as altogether constituted that they excite first the image of their union, then this union itself, but on the contrary it is the phantasy of their conjunction, sustained by their erectile capability, which gives them their form. It is for this reason that they belong to no one. Lacan has emphasized this for the phallus on semantic grounds, and Claude

Simon, in *Flanders Road*, from simple experience; one might say much the same for the vagina. The sexual organs do not bring about the Conjunction, for in that case they would belong to the partners. Once they have been formed by the conjunctive phantasy, they are the Conjunction itself – perceptual, imaginary, and motoric – in the course of self-accomplishment. Accordingly, there is neither attraction nor the *need* of a feminine pattern by a masculine one, nor the reverse, but *desire* of the copulation.

We have seen that the symbol supposes the signs, inasmuch as surpassing has no meaning save in relation to the surpassed. To arrive at a full comprehension of coitus, it thus becomes necessary to see how it connects with everyday life. In itself alone the differentiation of the sexes is a sheer superabundance of unity, and not yet, properly speaking, a determination. But as involved in daily life it is genuinely determined.

For the human being there is, in effect, a simultaneous appeal to two opposing orientations: to favor the discontinuous and live as a subject confronting the world, modifying it by work in a dynamic transformation; or to live as a subject-object united with the world, favoring its fresh growth and its own being in a dynamic adaptation responsive to continuities. It is impossible for the individual to accomplish these two aspects of his destiny to the same degree; he must favor one of them to secure inner coherence. Now the masculine genital form, convex and discontinuous, is more conducive to the first attitude, while the feminine genital form, concave and continuous, is more conducive to the second. From which, whatever be the cultural differences, there is an evident complementarity of existential styles between man and woman (Buytendijk), and a primary manner of being determined.

And, since the human being is a classifying animal, groups have taken up this distinction to construct systems all the way from astronomical phenomena to kinship ties, social stratifications, culinary prescriptions, technical terms. This is responsible for relating the masculine to the luminous, the solar, the mountainous, the arid, and the feminine to the shadowy, the lunar, the low-lying, the humid, with inversions, sometimes striking ones, of particular peoples. This procedure is very evident in cultures called primitive, but echoes of it can be found in the so-called evolved cultures, even at the very heart of scientific speculation, as Bachelard has shown.

It is thus necessary carefully to distinguish three levels: (a) the elementary distribution of the Conjunction as vital tenon and mortise, as it is experienced in the coitus itself; (b) the complementarity of styles of existence resulting from this distribution; (c) the concrete forms assumed by the existential styles in a particular cultural complex. The first of these levels is that of the symbol. The third is that of the signs. The second is the mediation through which the symbol acquires the signs whose negation comprises its meaning, and through which, in return, the signs guarantee remembrance of the originary bond within its diffused everyday forms. This intermediate level is close enough to the Conjunction so that through it the commonplace manifestations of the masculine and feminine assume, even in their artifice, the quality of ontological complementarity.

Moreover, the coitus is linked not only with the entire existence of the individual, but with his beyond, his issue, in the eventual fertilization. As continuation of life, insofar as it introduces a combination of genes which is an extension of the couple become visible, the fertilization opens the Conjunction (without breaking it) to a fresh proliferation of signs and operations. As moment of death, for it announces the relief of one generation by another, the fertilization summons the Conjunction, in extreme fashion, to the negation of every sign and every individual. The fertilization undoubtedly affords the ultimate distension of the symbolic.

The necessary articulation of sexual symbol in terms of signs, despite the inner logic of coitus and despite its fundamental intention, results in a variety of sexualities according to individuals and cultures. Thus the pervert, characterized by failure to attain symbolization, contracts the symbols into signs: his desire, instead of liberating in the joined bodies the infinity and immediacy of phantasy, shrivels to *objects* or *organs* (fetishism, voyeurism, exhibitionism, sadism, masochism), or encloses itself in *roles* anterior to the Conjunction (homosexuality). Without going as far, the West, whose merit it has been to abstract pure signs, has had to express the sexual act in the most operational vocabulary possible: these have been the reproductive and hygienic conceptions, reducing the sexual act to a *means* of propagating the species or securing physical equilibrium, or to the hedonist conception, which sees in it a simple pleasure. Such activism borders on perversion – a phenomenon essentially Western – but it customarily remains sufficiently theoretical so that it does not truly compromise the acts, salvaging in actuality the symbolic dimension repressed (or precluded) in the theory.

By contrast, non-Western peoples have always conceived coitus as symbol: in a cosmo-vital form in Africa and India, "erotic" in Greek esoterism, creationist in Israel, orgiastic in countless dissident groups. Then it is the signs, in the form of mythological proliferation, which menace the symbol, and in these groups sexual mythification assumes the role which perversion plays for us. Finally, in the countries of advanced industrial civilization, a new type of sexuality, which one could call interpersonal, is born. If elsewhere the partners obliterate their singularity for the sake of the Conjunction, here they underline it; and it is the personalized flesh, this recent invention, which, in its singularity, conjoins the determinism of the sign and the infinity of the symbol. Yet perversion is always at hand, and Sartre has described an interpersonal nuance of sadism and of masochism.

The evocation of signs by the symbol also explains the genetic development of sexuality, whose initial manifestations are to be found in the erections of the nursing infant, joined, like the earliest smiling, to the so-called REM sleep, in which dreaming occurs. If it is true that the erection already commences the desired Conjunction, if smiling is, as Freud thinks, an acquiescence not to the particular but to the world in general, if the dream provides the least restrictive liberation of fantasy, we have there the original nebula of sexual symbolism, and even of symbolism in general. Then, in the same way that learning seeks to record or to establish differentiated signs, it is possible to see a quest for self-regulation spurring the individual on to annex more and more extended and abstract realms. This is the viewpoint of the logicians like Piaget, for whom after all the truly real is to be found in the functioning of comprehensive signs. But this movement can also be that of symbolization, which, in order to grow, or simply to remain conscious, requires circumstanced mediations. A dialectic of this sort is particularly clear in the resolution of the Oedipus complex, where sexuality passes beyond the family circle in the direction of social preoccupations in the latency period, then toward the adult choice of a sexual partner outside the family circle, and that less by reason of an irrational defense forbidding the mother and father – except by historical accident – than by a requirement of the sexual intention itself, whose symbolic outreach would end in failure if it were to enclose itself in the semantics, so soon stagnant, of the parental triangle.

Thus it is necessary to be careful in deciding whether the sexual symbolization is archaic or creative. As a symbol it always refers back to the initial nebula. But this nebula, in order to continue to exist, in order not to lose consciousness, must differentiate itself from a differentiation, which each time it negates. Such an unfolding, always recognized in vain, is undoubtedly the very movement of existence, flux and reflux together. It is strictly the case that the more archaic the coitus is, the more it means the future. Just as art is the more creative as the primary fantasy is more liberated. As the mystical culmination is the more inspiring of action as its abandon is without return.

Of all experiences of symbolization, coitus is the most primitive. We have surmised its presence in the smiles and the precocious erections of the nursing infant during REM sleep, the sleep of dreaming. But theoretical reasons can be given, too. It is in the sexual experience that the articulation of the symbol into signs fundamentally occurs: the body of the subject contraposed to another body of the same species, the latter convoked by the erection (masculine or feminine), of itself conjunctive. Contrariwise art has for its material an object which is nothing but a quasi-subject (Mikel Dufrenne), and the mystic Self supposes the world. As for the act of symbolizing, since we have seen that it grounds the symbol, it assumes equally in coitus, at least as fantasied, the most elementary of forms: that of orgasm pursued for its own sake. Contrariwise, artistic creation or mystical ecstasy have recourse to neuron synchronizations of the orgastic type – without which no symbolization is possible – but *with* the subleties and moments of suspense that introduce into perception, imagination, and performance (as in the perceived, the imagined, and performed) the distance of knowledgeful mediations.

If it is not true that the sexual act is the model of all symbolism and all semantics, it remains their permanent root. Other levels of symbols, art and mysticism, other systems of signs, science, technics, logic, develop latent characters irreducible to the sexual act. But its priority, both temporally and dialectically, makes them lose their originary bond, so that they become rank or wasted – neuroses or psychoses – as soon as they neglect its renewal or its recall.

transl. E. E.

Source: *Cahiers internationaux de symbolisme*. Numeros 15/16, 1967-68. pp. 93-101.

MAURICE-JEAN LEFEBVE

The Nude as Symbol

"We do not make use of things," writes Senancour, "but of their images." Accordingly I would like at once to invite you to consider the nude as an image, and not as prey. That can at first appear difficult, but this difficulty ought not detain us any further than does the nude itself when it fascinates us. This difficulty ought to give us pause, in the same measure as the nude, which when it proffers itself to us, agreeing that we make some use of it, nevertheless prompts us to take dilatory measures in its regard, so that we postpone gratification in order better to be receptive of sheer fascination. Here I could quote Lautreamont when, at the end of one of his odes, he speaks of letting his inspiration gain its breath for a moment and then compares the poet to the lover who pauses in the midst of his act to contemplate his desire. Thus contemplation replaces action. The nude, at that very moment, is above all an image. In proposing that you consider it as an image I am presenting you at the same time with the thesis I wish to establish here. That is, that the fascinating power of the nude, the erotic fascination, whatever its physiological roots might be, is nevertheless to be interpreted and grasped as an imaginary phenomenon, an ideal phenomenon. There is, in the fascination of the nude, something other than an instinct which is self-triggered and seeks only its own end.

I have said "image"; I could have in effect said "symbol." Nevertheless the nude body gives itself to us, within reality, as a simple object. What is the difference between an object on the one hand and an image or a symbol on the other? We know that it lies principally in the use we make of it. The object, whether natural or artificial, has merely the role of a *means*. We make use of it to attain something else. The knife and fork are for eating, the armchair to sit on. Thus they are not ends in themselves, but simple relays, links in a sequence of actions, and we pass through them without thinking of them. In that respect any object is no different from the sign whose function is as fully utilitarian, since it disappears as soon as it has invested the mind with its designatum: the thought pursues the word. But we do not make use of a symbol as we do of an armchair or of a fork. The thought which envisages a symbol never entirely takes the place of the form which incarnates the thought. Differently stated, the symbol is not solely a means, it is an end as well: it is *the means to its own end*. It is not solely a mode of thought, it requires itself to be thought. And it is for that reason that it constitutes a problem. It indicates, but it indicates without truly revealing. While the sign puts us at once in possession of its meaning, the meaning of the symbol is always in question. The symbol is that which, in us, poses the question.

What question? The only one which preoccupies us for the moment is that which the nude poses for us. Before confronting this problem it is first incumbent on us to show that the nude is truly a symbol, that is to say, that it is not solely a means, but an end; not solely a prey, but also an image; not solely an object, but also the thought of that object. Before we inquire about what it is that the nude questions, let us show then that it is, in effect, a question.

II

Here a variety of evidence could be brought. First and foremost the fact that the nude and eroticism constitute a problem. We would not be preoccupied with knowing what problem the nude poses, if it did not in effect appear to us as a question. And the very word "eroticism" stirs up in our mind and sensibility vibrations at once penetrating and mysterious. The fact of eating, of feeding, does not represent a problem, insofar as the instinct of self-preservation gives us that act as an immediate means, attaining its end directly. But the amorous act does not as clearly indicate its end, i.e., reproduction, and "the perpetuation of the species" opens up quite different horizons than "the preservation of the individual." Outside of this evidence in some sense implicit, I shall offer three other arguments. The first concerns the sense itself of the word "the nude"; the second consists in showing that the desire aroused by the nude body does not find in it its exact response; the third, that the sexual act is itself a symbol for which one has always sought mythical interpretations.

An erotic undertaking customarily begins with sight. It is also possible for the spectacle to exist by itself, without being followed by any sort of outcome. The word "nude" has been borrowed from the vocabulary of the plastic arts. In sculpture or in painting the naked body of man or of woman has always (from cave art on) been represented in the same way as other objects. The nude is thus not the object itself, but its representation: it belongs to art, to the imaginary. From this one could draw the conclusion that the naked body which the artist takes as model is an object like the others, like all those which he has been able to represent at the same time on the canvas: the chair, the window, the apples in their dish, the cat by the fire. In other words the naked body would not be there, any more than the chair, the apples, or the fire for the use one could make of them in reality, but as a simple pretext for harmonies of line and color. In effect, the nude plays just such an altogether plastic role on canvases or in sculpture. Nevertheless it goes without saying that the fact of representing such an object in a civilization where it is generally hidden from view (whereas one would never dream of concealing apples or chairs, nor does one clothe cats or dogs), this simple fact implies that the naked body possesses a function in art which is different from that of the other objects. And whence does it receive this function, if not precisely from its turbulent part, from its erotic potentiality?

Moreover, we know that the problem has a long history, as evidence the prohibitions in Islamic or Japanese art, or as we see from the excuses and justifications advanced by painters and sculptors. They would have it that their art possesses precisely the gift of neutralizing the unhealthy and perverse part of the object represented, leaving for our contemplation only the aspect that is harmonious, tranquil, and elevating. Alain, in his "System of the Fine Arts" runs into this difficulty and gets out of it with a witticism. "It is commonly said," he writes, "that the nude is always chaste, if it but be beautiful; yet it would be better to say that the nude is beautiful, if it but be chaste." This is the giveaway. If I try to understand Alain's thought, I find only this, that he puts morals ahead of art, and that for him the only nudes which are able to have an esthetic value are those which disguise their erotic meaning beneath another, under a cloak which the moralist has cut to measure. There is no denying that the nude, thus accommodated, can become symbolical of all kinds of things. It stands for health, vitality, grace or even modesty. The nineteenth century abused these allegories in which luxuriant nudities take the liberty of representing almost anything one wishes: Truth, Liberty, Justice, Hope, not to mention Industry and Agriculture. But who doesn't see the artifice here, and how the idea of Alain overlooks one whole part, and indeed the most important part, of this kind of art? I am quite willing for the nudes of Rubens to be clarion calls of vigor and health, but it would be idle to deny that it is a matter above all of a vigor and a health which are extremely carnal. By contrast the nudes of Cranach offer us a morbid symbolism: these emaciated and wan bodies which appear stigmatized beneath the vanity of delicate black lace speak to us so eloquently of Sin, of the Fall of Man, of knowledge and of consciousness, just because they insinuate at the same time an evident sexual perversity. Neither the nudes of Goya, nor those of Ingres, nor those of Modigliani (to take random examples) nor even those of African art or of Hindu or Cambodian bas-reliefs can be characterized as chaste. And it is said that they do not belong to art! The Venus of Milo puts one in mind only of a linear maternal harmony; the Venus of Cyrene, for its part, if a symbol, signifies sheer eroticism. It is, from head to foot, aphroditic.

Nevertheless, it remains art, i.e. representation, image, symbol. But art invents nothing which is not already in nature. The transposition of art is required to endow a shoe, a chair, with esthetic value. The naked body has the advantage over other objects of already being esthetic in itself. Thus George Bataille, in his work "Eroticism," maintains that it is the more beautiful as its physiological functions are less marked: which is not to say, as in the remark of Alain, that its beauty is not of an erotic nature, but rather that the eroticism and the esthetic proper to it are not necessarily functional, that love and procreation are distinct. It is of itself that the body gives itself first to be contemplated before being possessed. And we will try to show (which is our second argument) that its possession alone does not extinguish the need which it announces. It remains a problem, even when one has tried to wrest from it the response which it promised. For that I turn to a variety of evidences, which I will relate to you briefly.

III

Here the series is, in effect, interminable, from Lucretius to Gaston Bachelard. They all come to a conclusion which can be very simply expressed: love (physical love) is a deceiver; it does not hold to what it promised. A pessimistic conclusion. However one can discover an optimistic aspect in it, and say for example with André Breton that "Woman is the great promise, which continues after having been held." The idea remains the same, and it is precisely that which we wish to clarify here, namely: that the object of love conjures significance, emanates an appeal which exceeds the use one makes of it, and which is never entirely fulfilled by it. It is as if the satisfaction of desire, failing to coincide with that desire, revealed at that very moment to a man, that he desired something else than satisfaction and, as Gide has said, that the importance was thus in his look rather than in what was looked at. From which point the man can, in effect, take himself to have been duped, or on the contrary, like Breton and like Gide, find in this discovery a reason to begin the experience again. It is with difficulty that one resigns oneself to leaving without solution a secret which lasts as long as one's life.

You may recall the admirable warning of Lucretius:

> There's the hope, always, that the fire may die
> Extinguished by the body which aroused
> Its ardor in the first place. What could be
> More contrary to nature? Nothing else
> Inflames us, once we have it, with desire
> Of more and more and more. We satisfy
> Our hunger and our thirst with bread and wine
> Whose particles have substance and can fill
> Our need, can take a place within our bodies;
> But pretty faces, fair complexions, bring
> Nothing to body's emptiness but only
> Frail, vain, elusive images, which hope
> Grasps for in vain across the empty air.
> In dreams, a thirsting man attempts to drink,
> But finds no water which can cool the fire
> Within him; therefore, all unsatisfied,
> Seeks images of water, comes at last
> To a rushing river, where he seems to drink
> In vain, still thirsty. So it is in love.
> Venus plays tricks on lovers with her game
> Of images which never satisfy.
> Looking at bodies fills no vital need
> However nakedly the lovers gaze,
> However much their hands go wandering
> And still are empty – can they gather bloom
> From tender limbs? And then the time arrives
> When their embraces join, and they delight
> In the full flower of love, or almost do,
> Anticipating rapture soon to come,
> The moment of the sowing. Eagerly
> They press their bodies close, join lips and tongues,

> Their breath comes faster, faster. All in vain,
> For they can gather nothing, they cannot
> Effect real penetration, be absorbed
> Body in body, utterly. They seem
> To want to do just this.[1]

A love code of the twelfth century, article twenty-seven, decrees in a
quite different spirit: "The lover never has enough of the enjoyment of
the beloved." This injunction is not difficult to respect, if one goes
along with Gaston Bachelard who rediscovers, in order to speak about
love, the Lucretian opposition of appetite and libido. "The appetite is
more brutal, but the libido is more patient," we read in *The Formation
of the Scientific Spirit*. "The appetite is immediate; for the libido, on the
contrary, are the long thoughts, the long term projects, and patience. A
lover can be patient as a scholar. Appetite is extinguished in a full
stomach. The libido is scarcely gratified before it is reborn. It *wants*
duration. It *is* duration. To all in us that *endures* is attached, directly or
indirectly, the libido." There is not a shadow of pessimism in this conception of love as research and conquest, ardent and painstaking
alike. On the other hand, the deprecatory viewpoint is underscored in
many writers. Pierre Bayle, at the end of the seventeenth century,
holds that "all pleasure is negative, all suffering positive." Love, he
adds, is merely a snare, a trick of the Creator to compel us to perpetu-
ate an accursed race, our own. Swedenborg makes fornication the
punishment (indeed, a bit charitably) of the damned of his Hell. The
image is picked up again by Baudelaire in his *The Women Damned* and
many other poems:

> Bent over his beloved the panting lover
> Has the air of a dying man, caressing his tomb,

just as in *Rockets III* he compares at length love to torture by electricity.
All in all, there are not a few thinkers ("intense lovers and rigorous
scholars") who, reflecting on the carnal act, have seen in it an obscure meaning which exceeds that with which physiology is satisfied:
at once psychological, social, metaphysical, religious. Jean-Paul
Sartre, for example, sees in it a kind of battle of the sexes, the means of
making the other a simple object in fixating his consciousness in the
denseness of the flesh. Keats announces that "the love of the senses is
based on the hatred of spirit." As if the libido itself had a need to be

[1] Lucretius, *The Way Things Are*. Transl. by Rolfe Humphries. Bloomington and London:
Indiana University Press, 1968, pp. 150-151.

psychoanalyzed. For the rest, all the Christian West, all the Buddhist Orient denounce physical love as a deception of Satan, a falsification of the divine Love, the shortest path to perdition. As proof they offer the fact that this love and its object, the naked body, are essentially material and gross, completely opposed to the spirit. Nevertheless, I believe that Lucretius was better inspired when he attributed the lack and the deception to the – shall we say spiritual – nature, in any case imaginary character of the nude. An empty form, he emphasizes, whose content we seek in vain. Isn't it mad to wish to possess an image, a simulacrum i.e. if we refer to Book Four of *De nature rerum*, these kinds of virtually immaterial layers constantly detached from objects to impinge on our senses? These are thus able, even if our eyes are closed, to cross the mind and rouse in us the dreams of sleep. That one is quite capable of making love while dreaming, isn't this the incontrovertible proof that it is nothing but vanity? Suppose, on the other hand, that it is a matter of a direct image, that the body is quite present; then a veritable fury is unleashed in us. "The body seeks the source of the mind's wound," writes Lucretius, "All things tend woundward: does not the blood spurt out/In the direction of the blow?"

Love and Death, the theme is old as the world. Here is their struggle in the agony where the sufferer wishes at least in dying to wrest its secret from what is killing him. For it is indeed a question of a secret: the naked body is now no longer the simple and naive response to the need which it excites; it represents for us the unknown. Shadowed or radiant, it is a riddle through and through. I would like to quote a beautiful and very revealing line from Theodore de Banville. The author evokes for us the strange elysia of which he dreams, and then he offers us this promise:

> Our eyes will become drunken with feminine forms
> More beautiful than bodies.

We see that it is a matter here of being satisfied with, of feeding on, not the material part, but the imaginary and symbolic part of the body. There is no form in existence which does not require its meaning. The beauty of the naked body is not only in the harmony of its lines: the fullness of Rubens or the morbidness of Cranach can be equally disturbing. What is then the sense of this nude now altogether formal? What is the meaning of the act, of which it appears to be more the pretext than the object? You know that answers have been found. They are the myths of love.

IV

I will try to relate them briefly. But in doing so, I will be exposing my-self to a number of difficult problems which, it goes without saying, I shall not be able to answer. I ask to be excused. For the rest, all that concerns me here is to show that sexuality and the nude have given rise to various interpretations, that meanings have been claimed for them which exceed their simple physiological nature, and finally, that they have generally been considered as signs or symbols.

I will only mention for the record the Pythagorean myth which dis-covers in the figure and proportions of the body examples of the mysticism of numbers, live applications of the laws of symmetry and harmony. We know how, from antiquity to the Renaissance, this theory has held sway over the arts, in particular, architecture. The human body appears as the perfect module, whose proportions are in agreement with the Golden Number, serving to regulate the most various works. Vitruvius, in his *De Architectura*, shows that the propor-tions of the temple ought to reflect those of the human body. The metaphysics underlying this esthetics is that of the analogy between the microcosm and macrocosm. If the body is beautiful, it is not only that it is harmonious, it is because this harmony is the very same one which governs and conjoins the different parts of the universe. The body appears at the same time as symbol, double, and epitome of the world in its entirety.

The Pythagorean man-microcosm is asexual. This brings us to another myth which has more interest for us: that of the original androgyne. Plato's *Symposium* tells us what happened to these first hermaphroditic beings who found themselves one fine day divided into two, and who ever since have sought to rejoin one another in love in order to reconstitute their lost totality. The Platonic fable is, as we know, merely one of many, many examples we have of this story. It is to be found in the *Upanishads* as in the *Old Testament* (Eve created from a rib of Adam), in Iranian or Greek mythologies as in the cults of African tribes, among the Christian mystics and in the symbols of alchemy. Mme. Suzanne Lilar, in a recent book, *The Couple*, has given an excellent summary of the different forms of this myth. She finds a confirmation of its importance in the Jungian theory of the *anima*, as well as in the most recent discoveries of genetics. What we have is a myth of reintegration and of Paradise Lost, where the original postul-ated androgyne is, in short, symbolic of the primordial unity. Why this

unity is discovered to have been lost is for other myths, such as those of
The Fall and Sin, to explain to us. Accordingly, ever since then, *illo
tempore*, man has been separated from himself in the person of the
woman, and hence the lovers wandering in quest, each through the
other, of their primal integrity. Love and desire are nothing but the
nostalgia of a perfect state in which there is no place at all for love. To
this central myth are attached, as Jung has shown, the alchemical
myths which extend the desire of reintegration to all of nature, or certain psychoanalytic interpretations like the birth trauma studied by Otto
Rank. In the latter case, the metaphysical nostalgia we ascribe to love
would be merely the sublimation of a purely physiological longing for
the maternal breast and the soothing loss of consciousness it brought us.

Love in the West, sentimental love, courtly love, passionate love is,
as we know, an invention of the twelfth century troubadours. Its
origins are social (since it reproduces the feudal structure in amorous
relationships) as well as religious. In general it is, as in *Tristan et
Iseult*, an adulterous love. For that reason it is condemned never to
be able to be completely satisfied; by the same token also it becomes
the symbol of another quest and of another adventure: the relations of
the knight and his lady are none other than the image of those of the
soul and its God. The Grail, filled with the blood of Christ and symbol
of redemption, could very well be considered as an erotic object. Thus
we find again and again, in the Western conception, from courtly love
to the Eternal Feminine of Goethe, the same nostalgia of a reintegration with what one can term Unity, God, Mother, or Nature. But the
name is unimportant. *In all cases, love and the sexual act are conceived as
symbols, in the very measure that, deluding themselves in some kind of object,
failing the end for which they appear to have been disposed and articulated,
they appear as a form whose intentions are not fulfilled in reality, so that nothing
is left to us but to imagine a content for which we are to seek elsewhere the
model*: in the depths of nature or in that of the origins, in primordial
simplicity or in the final and complete union with God.

V

"In the beginning was the *fabula*," wrote Valery, "so it will always be."
Just now I said that in recalling the myths of love I would touch on
difficult problems, perhaps insoluble ones. What credit to give, for
example, to the myth of the original androgyne? What is original is by
definition out of reach. Furthermore, to say that sexual desire seeks its

assuagement because the conscious duality yearns for the unconscious unity has a marked resemblance to a tautology masked by a trick of vocabulary.

Nevertheless the myths have their importance. What is important, is not the answer they bring or believe to bring, it is the simple fact that they underline, by their very existence, the problematic character of eroticism and love. The sexual instinct is something other than an instinct, the sexual act is not an act which achieves its end precisely, the sexual object is not an object like those of which we make daily use without any thought for them. Here is the gist, the central fact we have been able to disengage and establish. In the latter part of this communication I would like to ask you to put aside any interpretations of this fact, the answers one brings to the question, so that we can inquire simply about the question itself, about the form of this question, about its nature as question. And here I shall come back to what I said at the beginning, that is, that the nude gives itself to us as an imaginary object. I would like, if not to establish, at least to suggest this thesis: that the nude is an image, and that the desire and the pleasure of the senses are not essentially different from those which other images communicate to us, for example, those of art.

As is known, a certain current of psychology has claimed to trace back the charm of images, those of the dream for example (but those of works of art as well), to the occult and generally disavowed effects of the libido. That is, every harmony and fascination, the most beautiful spots in nature, the purest movements of the Greek tragedies, the sparest lines of temples, would be finally reducible to eroticism. I would like to demonstrate virtually the opposite, i.e., that it is the appearances and thrusts of love which should be interpreted from the symbol, to show at the very least that purely esthetic beauty on the one hand and erotic beauty on the other, have fundamentally the same structure.

And we have already shown this, in essence, in underlining the symbolic scope of the nude. For in both cases (esthetic and erotic) it is a matter of symbols or of images, that is to say, objects which designate something other than themselves, which pose a question, and pose it without at the same time bringing the answer which could extinguish it. Every esthetic image (whether it be a fact of art or one of nature) is a disclosure which eludes, an imminence which finds itself instantly deferred. What inquiry at once patient and suffering do we pursue through symbols and images, whether those of art, of nature, of space or of time, of dream or of love – what inquiry if not that of *Being*? The image is this sign whose meaning slips away at the very moment it was to have been delivered to us, so that there is nothing else for us to do but invent it, or, better yet, for us *to make* ourselves this meaning, *to be* this put-off promise. The image is that reality whose substantiality dissolves in the moment we near it, in such a way that nothing is left for us but to take the place of this void, to make up for this sudden lack of reality by what we can term a *surreality*. It is Being, finally, which is the sole profound content, at once immanent to all the symbols and absent from all the images. And, to be sure, I do not deceive myself that the word is just as vague as it is empty and contradictory. By no means do I pretend that by way of the images and symbols we really attain to that myth among all myths which I have ventured to term Being, which one could with equal justice (or injustice) term the divine, the sacred, or more simply, the mystery. I say simply that it is through them that we try to reach it, and that what we call esthetic pleasure, enthusiasm, or ecstasy, or love, is always the same rapt devotion to a sovereign Good, sovereign and radiant only because it is never more than promised.

Fundamentally, we do not contemplate a nude body any differently from the way we contemplate a landscape, a sunset, or a painting of Mondrian. Right away I hear you objecting that the sunset or the painting are content to be merely contemplated, while the body signifies a passage to acts. But what acts if, as Lucretius sensed, they are not just as many questions with which we surround, grapple, and assail the object from which we expect an answer? The difference, it goes without saying, is that the painting or statue or musical passage each claim but one of our senses, while love claims them all. Moreover it appears that in the last instance sensation is dominating and superabundant to the point of no longer leaving any place for reflection, for consciousness. But one also knows that in a work or spectacle to which one is truly wed, the very body of the material used, the pastes of the painting, the fleshly charm of the words and sounding rhythms of the poem have a sensual impact on us which makes them like caresses. I have not sought to play with words in using terms like wed, body, fleshly, and caresses: they offer themselves as the most appropriate terms when we want to speak of a work with love. As proof, this phrase of Valery in *Fragments on Art*, which could relate to our remarks when it seeks to define the order of things esthetic: "In this order," he writes, "*satisfaction* reawakens the *need*, *response* renews the *demand*, *presence* gives rise to *absence*, and *possession* to *desire*."

VI

The contrary is moreover equally true; that is to say, if one attempts to give an idea of the nude by means of language, recourse must be had to all comparisons and all metaphors, as if it were necessary, to unveil it ever so little, to invest this strange and fleeting form with every aspect of the world. On the one hand, the great natural spectacles make us think of the body (which we have noted in recalling the myth of man as microcosm); on the other hand, the body makes us think of all the aspects of nature or human art, as if we, in effect, rediscovered in it those illusions which can so unsettle us at special moments, the sphere at hand or far off, the trembling of the leaves, the ocean sparkling, or the sun which dazzles. ("Man," says Aristotle, "is begotten by man and by the sun as well.")

You will allow me then to conclude by simply evoking some images which give abundant proof of what I have proposed. Thus it is easy to recognize what the poet of the Shulamite is saying to us when he evokes the frolicking twin does, the herd of goats on the side of Mount Gilead and feeding among the lilies, the pillars of marble set on plinths of purest gold, the secret garden, the wellspring and the hidden fountain. But Saint-John Perse, in *Seamarks*, uses images borrowed from navigation:

". . . And what is this body itself, save image and form of the ship – nacelle and hull, and votive vessel, even to its median opening; formed in the shape of a hull, and fashioned on its curves, bending the double arch of ivory to the will of seaborn curves. . ."[2]

One grasps the force of these metaphors if one remembers that the nude is an image. This means that it is also its own image, ever a bit apart from itself, resembling rather its own simulacrum, or as if it were always reflected in a surprising mirror. To make use of comparisons in place of it is thus to represent it by showing not the resemblance but the difference, the difference from itself and the aspects with which one overloads it. It is because it does not keep its promise, or, as Breton says, because it remains promise even afterwards, it is by this that it distinguishes itself from ordinary objects, those which one terms "usual." Because we use them as means to attain a given end, it is for this that it becomes its own end and suddenly opens into the imaginary. It is at once the place of possession and the beyond of possession; this act, Proust says, we take for a reality up to the very

² St. John Perse. *Seamarks*. Transl. by Wallace Fowlie. New York: Harper Torchbooks, 1961, p. 109.

moment we wish to make use of it. It is at that instant, that fleeting instant which undeceives us, that it exercises its fascination. The nude as nude is a *trompe l'oeil* which goes on and on: which deceives and undeceives us at the same time and continually. If woman is the great image, who remains an image even after one has unveiled her, it is because she has never ceased to pose for us the dilemma which so tormented Baudelaire: she is as incomprehensible as the divine. But the divine, says the poet, is not incomprehensible except that its transcendance blinds us, while it is altogether possible, he adds irreverently, that woman is incomprehensible because she is perfectly empty and has nothing to reveal. The metaphors we have cited earlier are static. Now, if the nude is an image, if by that very fact it promises and defers its secret at the same time, we see that its enigmatic power is better conveyed to us by showing it in its successive appearances and disappearances, always ready to be born from itself and to escape from itself. Every great image undergoes the metamorphoses of styles, and many have succeeded one another, from the *Kama Sutra* right up to our present striptease. Spenser, for example, describing the fairy Acrasia, tells us that

> Upon a bed of roses she was layd,
> As faint through heat, or dight to pleasant sin
> And was arayd, or rather disarayd,
> All in a vele of silke and silver thin.[3]

The very same writer has his hero Guyon stop by the side of a fountain where two young maidens are bathing. They are bathing, or rather struggling, each apparently taking turns holding the other under the crystal surface, until suddenly the first one rises up, as far as her middle, to be followed by the second whose heavy blond tresses, suddenly falling down, veil her again. Once more Guyon sees them from without, while he had plunged them into that water in which the bathers of Bern, to refer to Casanova, perceived nude young maidens who were there for their service. All this puts us in mind of the gardens of Armida, or even that picture of the creation of the world which Hesiod paints for us, and where we see Venus rising from the waves, still covered with a veil of foam. Then, says the poet, all the scattered movements impressed upon the universe by love directed themselves toward her. But the same image can make itself more subtle and refine even further the metamorphoses of which the nude is capable. Thus Hein-

---

[3] *Faerie Queene*, II, xii, 77.

rich von Ofterdingen, dreaming of swimming in a subterranean lake, each moment feels his breasts and limbs rubbing against feminine forms as evanescent and heady as a perfume which dies at the moment it makes drunken. Which makes him speak of the "dissolved maiden," and Bachelard, who comments on the passage, writes of the "maiden in the partitive case."

Inversely, it suffices for the nude to show itself full face, like a butterfly mounted on a piece of cork, for it to appear as the riddle it is, in the retreat and dissimulation proper to it. The nude, then, withdraws beneath its own nudity, disguises itself. And it is hard to keep from thinking that one is dealing with a mask. A mask: that which makes us, properly speaking, gift of a second sight, since the mask is nothing but the appearance of a reality which resembles it (the face), and yet is infinitely unlike it, every mask suggesting, at the limit, that it does not truly hide anything except an absence of face. It is actually that which accounts for its hallucinatory character. The nude appears then as a face signifying nothing, having nothing but an appearance of expression, a simulacrum of regard. Such is the strange waking dream on which some poets or some painters have fed. Those, for example, who delight in animating the automata, the most sensual signs of life. Saint-John Perse again, when he speaks of the "longhaired mask of sex." René Magritte finally, has conceived a body of a headless woman with a massive head of hair, so that tips of the breasts form lifeless eyes, the navel an embryonic nose, the sex a mouth with oddly pinched lips. And so are negated, in this perplexing visage, the meaning and intention which these traits ordinarily lend to each face. It would not be difficult, starting from there, to imagine the attraction capable of being exercised by the nude and lifeless body, more precisely, decapitated. There is nothing more poetic and more erotic at the same time than a young woman dead and denuded. This necrophilia, as we know, is not foreign to the romantic extremism of Sade, of Goya or of Edgar Allan Poe.

The nude is able to frighten as much as it attracts, and in the same instant. That is an ambiguity which it shares with all the symbols whose ambivalence is well known, identical to that of the religious or sacred. Finally I would like to evoke a last analogy. The nude, on the one hand, the images, all images on the other, have this is common, that they give rise to adoration as well as contempt. The nude has its idolators as it has its iconoclasts. "The common folk have been led into error by the poets," writes Bhartrihari, "and knowing full well that the body of the beauties is made of skin, of flesh and of bones, they render it a mysterious cult." Thus it is that there are the devotees of Love, and others who challenge their devotion and its object. Love wavers from platonism to sadism. For the nude symbol can be considered as much a means as an obstacle. It would need a book for itself to list the variety of experienced values which, from Plato to Sade, from the Bible to Baudelaire, have been attributed to love, to woman, and to her body. It is apparent, for example, that at the end of the eighteenth century, sadism and the guillotine conveyed the same dizzying irreality as the cult of Dionysos and his maidens, who in Phrygia were flagellated before the altars of the god. This problem has been illustrated by most beautiful images in Pierre Klossowski's "Diana's Bath." Here he shows us Acteon, first, to be sure, attacked by the goddess' hunting-pack (for, surprising her while bathing, he caused her loss of shame, without which the gods are unable to recognize one another). Acteon turns in sudden revolt against Diana and flogs her at length with the silver crescent (lunar symbol) he has torn from her, filling the countryside with his insults and his imprecations:

"Shameless bitch! he cries at her."

But Diana does not deceive; she is the simple illusion of stone. And all this violence is lost among the echoing woods under the vault of evening where the new moon is already rising. The goddess has assumed the prerogative of all divinites. She has regained the heavens, inviolate despite our modern rockets, from which she turns to us but one simple face, so that we are never able to tell if we are seeing her real face or only her mask – in other words, whether we are beholding the illusion of the being or the being of an illusion. What good to shatter this distant haunt, if she herself gives us the proof that her disappearances serve only for her to be reborn? For it appears that the revolt against the images is nothing but another way for us to pay them the absurd and inescapable worship due them, the cult which we have found brought together today under the joined signs of these two imaginary truths: Symbolism and Love.

transl. E. E.

Source: *Cahiers internationaux de symbolisme* Numero 11, 1966, pp. 3-15.

F. J. Smith

Don Juan: Idealist and Sensualist

It was Kierkegaard who gave us the modern concept of *Angst* and its full spectrum. We say *Angst* rather than "anxiety," since as a term the latter is far too psychological a notion for our present purposes. Anxiety is only one facet of the *Angst* spectrum. The whole spectrum takes in everything from that wonder, which is the beginning of philosophy, through adventuresomeness to alarm, and that anxiety born of a consideration of the possibilities a new adventure brings with it.[1] The word *Angst* (in Danish, *angest*) is a body word, not a lexicographical item useful to psychological science. Some of the original "body meaning" comes through in the modern word. Anxiety is not just a psychological state but a bodily "stifling," an inability to catch one's breath (*ango*, ἄγχω). In his Don Juan essay Kierkegaard rediscovers anxiety as musical and erotic.[2] In Christian and post-Christian civilization sensual-eroticism is born in anxiety; it is eros caricatured, caught between the extremes of idealism and sensualism, as personified in the tragic figure of Don Giovanni and in the sensuous music of W. A. Mozart.[3] Don Giovanni feels the stifling effects of the classic metaphysical dualism: idealism vs. sensualism. And Mozart effectively portrays it in his music. It is not that – to put it facetiously – Don Juan is "out of breath" because of his sexual exploits. Rather, he feels this stifling at the very outset of his amorous adventures; and since he never seems to overcome the metaphysical dichotomy, his distress accompanies him in every erotic feat and makes it impossible for him to find love for all his amorous adventures.

---

[1] *The Concept of Anxiety* (Princeton, 1957). tr. W. Lowrie.
[2] "Die Stadien des Unmittelbar-Erotischen oder das Musikalish-Erotische" in *Entweder/Oder*, II, tr. Ch. Schrempf (Leipzig, 1939), pp. 33 f.
[3] *Neue Mozart Ausgabe*, II, 5, 17 (Kassel, 1968), *pass.*

1. THE *Angst* SPECTRUM: FROM OPENNESS TO CLOSURE

The word, anxiety, contains the original historical meaning, even though a transition has been made from the bodily to the psychological dimension. In giving etymological derivatives and writing of historical meaning I do not intend to restore what is now lost to us. I would simply fix the blame for the loss, not so much on time as past (and thus the problem of past meaning) as on our present alienation from the bodily significance of the words we use. In our efforts to regain lost terrain and to return once more to what Merleau-Ponty calls the *corps ontologique*, we may be allowed to make some use of historical meanings as a model.[4] One also bears witness to the fact that historical meaning has survived, however minimally, even in the alienation of the psychological sciences from bodily reality; thus in the "state of anxiety" some of the body word perdures, in that anxiety is characterized as distress which has "psychosomatic" implications.

When we say that the original meaning perdures we mean that it still originates meaning: what the word portrayed then still happens now. The original experience is one we can still identify with.[5] We, like the ancients and our forebears, still know what a stifling feeling is, even though today we say this is the result of a "state" of anxiety. To cut off original experiential meaning and settle only for the "present" meaning of words is to be guilty of a historicism of the present. One tries to escape one's roots. One becomes a tumbleweed, anchorless, traditionless, abstract. A historicism of the present is even worse than a historicism of the past, which recognizes only old meanings and ends up in a kind of archaic romanticism. Word meaning, as existential, i.e., as covering human existence in all its facets, thus also in its past, present, and future, embraces the entire extent of history. We are still that which we and our forebears have been; their bodily movement

---

[4] *La phénoménologie de la perception* (Paris, 1945).

[5] E. W. Straus, "The upright Posture" in M. Natanson, ed., *Essays in Phenomenology* (Nijhoff, 1966), p. 184, footnote 23. Here the author states that "the history of a word represents the sedimentation of general psychological experience..." "Sedimentation" is used here in a positive sense, i.e., as the accumulation of the wisdom of the human race, based on its experiences and held as such in the history of a given word. Of course, sedimentation, as precisely that which phenomenology brackets in order to get at the ideality of words and concepts, can be taken in a negative sense as that which impedes our proper understanding of words. Actually a given word represents a long history of positive and negative forces that man has experienced. Modern man should be aware of this history in order properly to be able to evaluate the effect of language on expression and life, precisely as contemporary (rather than as non-historical, which words never can be except in the abstract.)

perdures, often strikingly, in us, so that it can be recognized as such by those who knew, e.g., my grandfather and me (though I myself never knew him). Thus also with words, which are the partial expression of that fully bodily word, that man as *logos* is. Word meaning embraces all of history not just a segment of it, i.e., the essentially undefinable "present." Augustine spoke of the past-in-the-present and the present-in-the-present. Existentialism, of course, lays a special emphasis on the thrust of the future and on creativity.

To be anxious means to be short of breath, to experience a kind of suffocating feeling and thus to be depressed. One yearns to gain second breath, to be freed from distress, to attain renewed life in a return to oneself, which frees one from ideological dichotomies like idealism and sensualism as also from metaphysical trichotomies like past, present, and future. One desires to return to one's own body, to become one with oneself. Being short of breath means being stifled in the "realistic" and factualist view of life, facts interpreted in terms of a given culture or collective society. Anxiety is engendered by our falling prey to the real. We are caught in its clutch, like some lithe creature caught in the jaws of a beast of prey and having its life choked out of it. If psychology is the study of psyche, it ought to take seriously the originative meanings of *psyche*, not as "mind" but as the *breath* of the body, thus as bodily movement. Breath is not just physical inhalation and exhalation. It is the whole bodily movement we experience when we breathe in and out; it is our fleshly being in the world of life. Today we realize this at the practical ecological level: breath and air pollution. But it is essentially an ontological problem. At times, independently of the ecological situation, it becomes difficult to breathe. We could not breathe properly in the stifling atmosphere of a religious or political collective. Eros and breath! And this is a question of more than metaphor. A collective, when self-contained and "filled up with itself" (*en soi*), identifies itself with reality. One needs the open air to breathe in.

Anxiety is an entire spectrum. We are used to thinking of it only as a psychological state. But the entire range of *Angst*, which Kierkegaard describes, shows us that a primordial wonder and awe, a sense of mystery and adventure, which only a child can grasp, are important factors. And thus, to be anxious can also have a more positive meaning than the one thus far described in this essay. Anxiety is not just a western sickness, as it has been characterized. In the presence of adventure we are apt to catch our breath – a kind of "stifling" in a more positive sense. We are rendered breathless and speechless before the wonder of "all that *is*," especially when it takes on the form of some spectacular natural phenomenon, such as a brilliant sky at night, an Alpine landscape, a symphonic composition, or that *surgissement* one experiences in eros-movement. For eros need not be merely erotic; it denotes rather the *surging forth* of our whole bodily being in joy and in love. This silent language of breathlessness is the prelude to "second breath" and the joyous acceptance of a new dimension of life, especially with another. This joy seems to have been denied Don Juan. Psychology has overemphasized the element of alarm in this catching-of-breath. As a science, it identifies with Don Juan himself. It is true, that if I stand before an abyss I feel vertigo coming on, and thus I catch my breath also in alarm, not just in awe. This vertigo is directly connected with that vortex of reality which we encounter at this precise ontological moment. The momentum of the ontological present carries us with it and thus effects that "dizziness' of which Kierkegaard wrote. This is the moment of crisis, i.e., of decision and parting of ways. In the movement of decision we have an "anxious moment." And this particularly when playing the role of Don Juan.

Stifling need not be interpreted only in a negative sense. If anxiety means catching one's breath at the possibility of existence at full tilt, to use a choice metaphor, it is something quite positive. Catching one's breath at the potential offered in the moment of existential decision also has the effect of making us aware of the stifling situation of a life led in a world of mere facts or in that ideological realism known to a false collective. It is true that there are certain types of people who seem able to live and even thrive on stale air, but whether this is the ideal man, religious or political, is a moot question. There are at least a few who need to breathe anew. They feel the capacity to breathe in the earthy atmosphere of freedom, this especially in matters of eros. It is this breathing anew that is the bodily basis of the "ontological" dimension, conceived as "letting *be*," allowing body to be bodily not merely physical. It connotes the overcoming of the stifling atmosphere and anxious breathing of those who are tightly bound to the merely factual order of everyday, a day caught between the extremes of idealism and sensualism. In this second breath or second spring we rediscover who we really *are* and begin the process of working our way out of dead traditions and forms. We rediscover eros, stripped of mere eroticism.

If anxiety has been described as a "western sickness," this seems true only from the psychologist's standpoint; for, *Angst*, as wonder, is hardly a disease, though it may render us a bit ill-at-ease on our "ontic" plateau, i.e., in our factualistic existence. If being ill-at-ease were a kind of disease, it would be best characterized not as an illness as much as an uneasiness. In its inception in modern times anxiety, in the writings of Kierkegaard, was a theological problem, one that re-echoed Augustine's "terror before the Holy" as well as Luther's theological pathology. In the case of Kierkegaard it seems reasonable to ask whether the theological dimension is not a kind of cover-up for his impotence vis-à-vis his fiancée. Kierkegaard backed off from confrontation into the easement of theology. Instead of following through with the adventure that beckoned, he eased away from it: from reality to books, from encounter to encapsulation in Christian thought. But even from his defeat Kierkegaard rescued himself and gave us the fruits of his struggle. Without Kierkegaard on anxiety, there could be no Heidegger or Sartre, at least not in the form we know them, as writers on being, time, nothingness, and anxiety.

Is anxiety, we may ask, a "natural" thing? And this especially with regard to eros? This might well be equivalent to asking whether being caught in the Don Juan syndrome is "natural." Can we be all that dogmatic in asserting that man, as such, always and at all times must have felt and has felt what we call anxiety, and that this is an absolute port of entry to the ontological dimension, as Kierkegaard and Heidegger seem to indicate? Such a dogmatic pose would seem to this writer somewhat hazardous. At most we can allow ourselves to speak of an anxiety idiom. This leaves the door open for a non-anxiety idiom or another approach to Being besides what we call dread. This writer would prefer to be cautious in suggesting that it is "man's nature" to experience dread and thus enter the kingdom of Kierkegaard's theologisms or Heidegger's ontologisms. The historical roots of this "anxiety idiom," which is transmitted to us through Kierkegaard, are to be found in the Judaeo-Christian culture, inasmuch as the Jewish and Christian covenants relate of relevant anxiety myths and folklore. The flood story, for example, is not an original creation of the Jewish Testament, as is well known by biblical scholars, whether Jewish or Christian. It was incorporated in the Jewish Scriptures from a common patrimony, not unlike the history of the musical instruments used in the Temple. The pre-Sumerian tale of the flood is repeated almost word for word in the Jewish Bible. Since man was becoming too "bold" to suit En-lil, the Storm God, man's annihilation by flood was decreed by this paranoid deity and summarily carried out. There was a gradual build-up of an attitude of fear and indebtedness ("guilt") in the ancient cultures and it culminated in an anxiety in the presence of the divinity.

The peoples threatened by the annihilation in the flood were confronted literally by nothingness and dread, themes so crucial to the works of Kierkegaard and Heidegger. The source of dread was the anger of the real or imaginary gods, whose mythology became an important tool in the hands of aristocracy or priest-caste in the control of people. The gods, and, of course, the god-kings, who were in authority as mouth-pieces of the deities, wielded a real power over their subjects in being able thus to threaten vengeance and induce anxieties. The feeling of guilt, as duty and indebtedness (*Schuld*) to the gods and to the king, spawned the dutiful good work. But this work was one of fear, performed under a cloud of anxiety in the face of impending annihilation, if the work were not properly performed and in due season. Thus also eros was turned into a "duty," one done in anxiety in the face of death. Even in detheologized form anxiety and the work of fear go hand in hand. Whenever we do not produce as much as we *ought* on a given day, we feel uneasy and somehow "guilty," as though we were serving some ancient taskmaster deity. The modern "work ethic" is traceable to the original slave ethic. We are still quite superstitious in this respect. Fear and guilt are the basis of many an institution's control over peoples' minds and hearts. Pure unselfconscious joy and spontaneous creativity are frowned on. The good man is the man of good works, whether this be done for merit or for pay.

Even the "everyday" of Heidegger's *SZ* can be thus regarded.[6] The everyday man is the working man, whether the proletariat or the performer of good works in a theological sense. He is the daily worker who works at producing so much product a month to fill such and such a quota, whether for a corporation or for a collective. A man does not even raise a family anymore; he produces offspring, whether for the City of God or for the State. Existence is a concretely work-a-day affair. The "ontic" order, i.e., the factual order in which one *is*, is one conceived in terms of a work-order, as a kind of absolute. If one does not work he feels useless and valueless, especially since he receives no monetary compensation for work performed. A whole world of values and evaluations rests on this basis, and the key to it all is fear and anxiety. (The ontological end of the *Angst*-spectrum is mostly sup-

---

[6] The abbreviation refers to Heidegger's epochal work, *Sein and Zeit* (1927) badly translated by Macquarrie-Robinson.

pressed or ignored.) The work-a-day world with its fears and manipul-
atable anxieties makes it possible for an assortment of monarchs and
deities, wearing either business suits or church raiment that would put
Solomon to shame, to control the working man. Since eros would go
against all this, it must be suppressed; and thus the ideal man is the
repressed ascetic or the dulled worker, turning out the proper quota of
products about which he cares nothing. Whole lives are wasted on this
treadmill, and these wasted lives and suppressed eros contribute to
monarchical or collective control by reigning powers. The fears and
anxieties thus spawned are in fact the major source of power for institutions. In all this true eros suffers. All that is left is eroticism as a
counter to idealism, i.e., to ideologies that build systems at the expense
of the drone population. For the drone's life is asexual. He is incapable
of eros; at best he achieves the erotic treadmill, like Don Juan. This is a
kind of compensation for the work treadmill he apparently cannot
escape.

For Kierkegaard fear is natural. He tells of the fearless young man
who had to go forth and learn fear; to be fearless is not natural. Yet
one can contest the interpretation of Kierkegaard particularly since he
witnesses to a theological tradition that roots in a historic conditioning
process in which fear and anxiety were major factors. Wonder, awe,
and mystery are indeed natural to man. Man is "naive," i.e., at one
with nature, thus he does not have any anxiety in encountering either
life or death. He had to learn dread. Christian civilization has stressed
mystery, it is true, but mostly in the service of anxiety and fear. It
has emphasized but one small area of the *Angst* spectrum, that area of
fear which might well be a natural concomitant of wonder but which
hardly belongs to its essence.

The non-anxiety idiom connotes the joyous acceptance of earthly life
and of death as part of it. In this joyous acceptance of bodily life in the
world and on the earth, one accepts eros minus the impositions of
metaphysics, thus eros without guilt, without idealizations ("Romanticism"), without that sensual-eroticism which was introduced into our
culture by Christianity. This is "naive" eros. Naiveté implies a certain
nativeness with reality ("nature"), a freedom and openness incomprehensible to those whose fears and anxieties shut them off from the
Good Earth and close them up within the hard shell of an ideology. If
we reject the historic anxietal metaphysics imposed on our culture we
have a chance to breathe anew, to gain "second breath," to recoup
human eros, as we return to our bodily self, freed of the strange gods of

history who preyed upon the lives and works of man. In an existence freed of theological dread and ontological anxiety, both impositions of history, man accepts himself and others joyously. In such a world no Absolute watches nervously over our lives, dictating what we ought to become; no institution imposes taxes on our joy; no false guilt complexes blight the authentic appreciation of life. To our culture, trapped as it is in its strange "order of reality," such a vision must be regarded as a utopian dream or as a kind of romantic anarchy. But it may well be simply insight into what man *is* minus the incrustations of philosophical theory and without the impositions of those gods who thrive on the anxiety of traditional man.

The joyful existence is not just a romantic dream or some impossible euphoric state. Rather, it is a Yes to life and an involvement in the *chiaroscuro* of existence. It is that reality which our culture has sought to suppress, because in the acceptance of joy and of eros all our societal forms would be changed. The gods of anxiety would be banished and man would be free to be himself. He would be free to wonder anew, to set out on a new adventure, free to live and love minus the fears and anxieties that plagued Don Juan and Kierkegaard himself. Modern psychology depends on historic sources for the concept of anxiety and it has lost sight of the full spectrum of *Angst*. The spectrum is restored to us by Kierkegaard. Psychology has fixated on the "ontic" end of the spectrum to such an extent that anxiety is made to represent *Angst* as such. As goes such misrepresentation so goes also any therapy that depends on it. It is to be wondered whether this might not adversely affect therapeutic sessions and give them a false emphasis. A patient might not be manifesting "anxiety" at all. Instead he might be experiencing a healthy case of *Angst* as ontological wonder. Thus he wonders why others are so closed to the dimension of joy, or the joy of eros, and even attempt to destroy them within themselves and in others. Perhaps it is from the frustrations of this typical contemporary situation that his protest "neurosis" stems. This is especially true when wonder and adventuresomeness (as an openness to new horizons of life and to a more truly human experience of eros) have been tabooed in a given institution, civil or religious. Our contemporary institutions are all faltering. Joy cannot long be suppressed.

In a collective society, in which various taboos have become a crucial part of the legal structure (in civil society pressure groups always seem intent on legislating their own hang-ups), the man who is indeed still open to the adventure of *being* a man is truly a "man in trouble." He is liable to be psychologized or demonized by the collective. A man who wonders in a wonderless society is liable to be regarded as "abnormal," particularly when the "norm" is the odd reality of a wonderless society. We can hardly expect to do much for someone suffering this kind of *Angst* by treating him as a case of anxiety. It is little use to the "radical" to regard him simplistically as a neurotic working at cross purposes with established society, when he is taken up with what went into society's becoming the strange behemoth it now is. What began in wonder and openness ends in closedness, what Kierkegaard has characterized as the "demonic" in man.

II. FROM KIERKEGAARD THROUGH TO MERLEAU-PONTY

The history of the concept of anxiety since Kierkegaard leads us through Heidegger and Sartre. It appears at least once in Merleau-Ponty, connected in that instance with the breakdown of language and its ontic chatter in the "limit-situation," which is the beginning of the "silent cogito."[7] The silence of the silent *cogito* is a silencing of ontic thought, i.e., of the accepted order of reality, of the accepted order of mental abstractions, and the opening up of the bodily thought of the concrete world. It is the "ontological" moment, i.e., the moment in which we are caught up in the movement of Being as such, rather than being trapped in a given order of beings. It is the moment when the limits of "normal realities" have been reached and seen to be but a dead-end. All the assurances of the ontic society are silenced, and the new creation is born in the dissolution of the old world together with its concepts, its language, its societal norms and taboos. The ontological dimension is the open area of Being as such. But it is difficult to verbalize about Being in languages that have been forced into ontic molds. It is particularly in the area of language that things have been closed off, and a veritable horde of border guards watches vigilantly over the boundaries of grammatical expression. And yet we need a revolution in language to match the contemporary revolution in thought and society. Perhaps Gertrude Stein was a witness to something important. In Merleau-Ponty the ontic chatter of words simply comes to a stop in the limit-situation. We become speechless in our anxiety. The order of reality and the abstract world of conceptuality breaks down completely, is silenced and dissolved, as we discover that now our body is engaged and "thinks for us," as it acts and reacts be-

---

[7] M. Merleau-Ponty, *op. cit.*, p. 462.

fore the mind can attend. The ideologies of the ontic order vanish. It is in the moment of silence that Don Juan could discover his true bodily movement and escape from the vicious circle of idealism-sensualism. This is why Don Juan *is* anxiety. He becomes the personification of *Angst* as silent dread. And yet he does not grasp the moment. It passes through his fingers. He moves quickly back into the syndrome that holds him captive, and all is lost. Once more he moves on to erotic conquest; and the brief vision of eros is but a dim memory, a dream from which he has awakened.

It is obvious that the writer is departing from the ideas of Kierkegaard as well as of Merleau-Ponty. It is evident, or should be, that this essay is an improvisation on such themes and not just an exposé of their thoughts. One statement from Kierkegaard's essay on Don Juan is of central importance to this improvisation: "Sensuality was introduced into civilization by Christianity." Don Juan is a type produced by Christianity, for he is the child of Christian ascetical idealism, the offspring of the monk, the personification of the sensual-erotic syndrome. The attainment of ontological body, i.e., body regarded as that human flesh and bone we already *are*, requires the overcoming of the tradition, which has taught us to look on the human body merely as physical or as physiological, as what remains after soul or mind is abstracted, as what the scientist regards as mere matter, eventually as what Descartes considered a corpse. And apparently even cadavers can be the object of eroticism in the deranged mind of the necrophiliac.

The "normal" man in our post-Christian culture is to a far greater extent than realized a Don Juan, torn between some Puritan ethic and his interest in pornography, between "true love" and the loved one's body as an erotic object. This is true of non-Christians as well, if we are to believe *Portnoy's Complaint.* The split between ideal and erotic love is portrayed in such films as *Bell' Antonio,* who idealized and thus would not touch his wife yet had no compunction about seeking erotic love elsewhere. Italy has produced the "*santa mamma*" who has a goodly number of offspring but drives her husband to a series of regular mistresses. Lands north of Italy have produced the matron priestess who knows how to castrate sons and frigidify daughters. It is strange indeed that woman has become the carrier of the primarily male ascetic virus, so that someone like Don Juan could be raised an idealist, only to turn into a devoted sensualist. It is strange that other religions, such as Judaism, could become so infected, thus going against their own originally healthy ethos.

The anxiety described by Kierkegaard in *The Concept of Anxiety* was most likely his own sexual anxiety, as he stood confronted with the possibility of marriage. The Don Juan essay in *Either/Or* is also no doubt a description of its author himself, thus making the essay "subjective." Anxiety cannot be treated objectively as ontological dread. The person of Kierkegaard is too involved in the concept itself; he does not stand by himself, however, even as an object of ridicule. He wrote as a Christian Don Juan. He fled from Regina Olsen to his writing table, where he put his anxiety on paper – much to our benefit. He became the object of that humorous adage he himself quotes: "Aut liberi aut libri." It is impossible to separate Kierkegaard's own erotic anxiety from the historical concept of *Angst* as such; and it may be important to note that the normal man's general anxiety roots more firmly in his sexual pathology than he might imagine, with or without Freud. We must bear this in mind particularly as students of Heidegger, for as we detheologize the Kierkegaardian concept we run the risk of de-eroticizing it as well. This can lead to a forgetfulness of the eros-ground of *Angst*, perhaps more important than what Heidegger calls the "forgetfulness of Being." Indeed Sartre is the first to point out that Heidegger's *Dasein* is essentially asexual. The ontological leap is apparently still ascetical, a mental exercise rather than a bodily spring, a "step back" but not a bodily return. *Angst* is not merely ontological wonder in an existential sense. It is erotic wonder primarily.

The phallic significance of existential philosophy is easy to see, whether we elaborate on ek-sistence in a bodily sense or on the more obscure "open middle" of which the middle Heidegger writes.[8] Heidegger makes much ado about the "potential to be" (*Seinkönnen*) and even derives art (*Kunst*) from primordial potency (*Können*).[9] It is bemusing to see scholastic minds reducing all this to a clear intellectual system, a system regarded as quite obscure from the analytic viewpoint. Ek-sistence is translated as a "standing-out." This can be interpreted in terms of man's learning to stand erect, thus making him stand out among all other creatures. It has obvious phallic significance as well, as does the wedding between the earth and the sky in Heidegger. Despite himself Heidegger has preserved eros as the ground of being. Ontological openness is not merely a philosophical adventure. Rather, it leads to the rediscovery of the ontological body, as Merleau-Ponty makes clear. The rediscovery of the eros-body means the silen-

---

[8] M. Heidegger, "Wozu Dichter?" in *Holzwege* (Frankfurt/M, 1957), p. 260.
[9] M. Heidegger, *Der Ursprung des Kunstwerkes* (Stuttgart, 1960).

cing of idealistic-sensualistic ideologies. The bodily *cogito* emerges in silence in that intersubjectivity which is incarnate in the act of love.

We have to redefine sex, inasmuch as it is central to eros and part of the problem of anxiety. To cope with sexual anxiety as personified in the figure of Don Juan, we have to discuss sensual eroticism as the antipole of ascetical idealism. The *corps propre*,[10] i.e., the human person as bodily rather than as a metaphysical abstraction or as the object of scientific study, emerges caricatured, i.e., in improper form, as the body erotic. This is not our "proper body," i.e., the body I experience as truly my very *own*, as that which I share with another, which I "owe" the other in marital "debt." My proper body is not erotic in the sense that it is an object for someone's detached fantasies. It is an eros-body, the love-body which is made to live with and within the other.

The body-proper, sometimes called body-subject, suppressed in Christian asceticism, emerges in its demonic aspects in the person of Don Juan, the erotic sensualist, the child of the ascetic. It was in fact an ascetical monk, Fray Gabriel Tellez, known in literature as Tirso de Molina, who reintroduced Don Juan in the Baroque Era. Tellez did not invent Don Juan, as a later Baroque music theorist, J. J. Fux, invented the "diabolus in musica"; for, Don Juan was the creation of the Middle Ages, as was the tritone in music. Don Juan was the incarnation and personification of the classic struggle between the "spirit" and the "flesh," with the emphasis on the flesh. The ascetic and the idealist must emphasize the flesh and caricature the body; for, the ascetic is in full flight from his own body and from any bodily sharing. The love-body is the source both of his anxiety and inversely of his ideology. Yet, as Sartre has pointed out, the more one flees from an object of anxiety, the more conscious one becomes of it.[11] The body-consciousness of the classic ascetic stands in direct proportion to his flight from the body as properly his own and as sharable with a potential loved one. The disembodied ascetic is condemned to continual re-embodiment. This is Don Juan's problem; for, since he cannot be at one with either his own body or with that which he shares, he needs many bodies. He measures his success in terms of numbers. But he never attains basic oneness, and thus all other numbers become meaningless, for all numbers follow upon the basic unit, one. The basic unit is the love-body which seeks union and communion with the other.

[10] M. Merleau-Ponty, *op. cit.*, p. 107.
[11] J. P. Sartre, *L'Être et le néant* (Paris, 1937).

Sexuality must be conceived in broader terms than genital sex as erotic. Even Freud did not conceive of sexuality as only genital, contrary to popular misconceptions. In phenomenological terms we might distinguish between primordial and physical sexuality. Primordial sexuality could be regarded as denoting sexuality *as such*, i.e., the human person *as* a sexual being. The Don Juan syndrome needs to be overcome. The idealist-sensualist complex blocks primordial eros; or, as we say in plain English, sexual hang-ups prevent sexual fulfillment. But sexual fulfillment, while including intercourse as its most intimate expression, also extends to life in general and to the arts and music in particular. It is for this reason that Mozart's music is "sensuous," especially in the *Don Giovanni*. Life in general and the arts in particular, if it is a question of human life and culture, are bodily. Art is not science. Eros is not merely erotic. Eros does not lead to moral dissoluteness, which is its own punishment. Mozart's title for the opera is "Il Dissoluto Punito ossia il Don Giovanni." The Don's life is its own punishment; for, "true love" ever eludes him as he conquers body after body in his vain and dissolute pursuit of a mirage. Yet Mozart conceived the opera as a *dramma giocoso*, as an *opera buffa*. And thus the humor of the situation is rescued. Donna Anna calls Giovanni "scelerato"; he retorts by dubbing her "sconsiliata"; and Mozart knew how to build this up until they shout the epithets at one another simultaneously (bars 97/8 and 115/16 of the score).[12] Thus Don Juan becomes a kind of roaring twenties' Casanova gangster and Donna Anna is mocked as a misguided virgin. (The opera could be re-done in contemporary form as a musical show.)

"Physical" sexuality is the product of metaphysics, if physical means that which is opposed to the spiritual or mental. But this opposition is a case of double vision: man viewed as a composite of two opposing factors. Yet man is not simply composed of "body" and "soul." Experientially, man is a living unity rather than the unification of two entities spawned in different realms. The Don Juan syndrome is that merry-go-ground (*syn-drome*), in which a man is torn between idealism and sensual-eroticism. Kierkegaard has written the classic essay. Don Juan is the incarnation of sensualism because he was raised in an atmosphere of ascetical idealism. He is the classic model of the good Christian man. From ascetical idealism to promiscuity: the story of the Christian Portnoy.

Man is not dualistic; he is indoctrinated to think so. He is rather an

[12] *NMA*, II, 5, 17, pp. 35, 37.

experiencing unit. He is not a unification of anything; rather, the whole body, Sartre writes, is psychic; the soul is the life-movement of the body and indistinguishable from it. One cannot abstract movement from a moving body except "in the mind." Sexuality must be understood not in terms of the classic dualism but as primordial. Man is not a "rational animal" but rather a living being (*zôon*) who experiences the world into which he is ingathered (*logos*). The experience is primordial, i.e., man has a basic and primal awareness of that world within which the external world of objects is but an abstraction. Primordial experience is an awakened awareness of that world we can share with another. It is an awakening in which we become aware of our truer nature, one not yet forced into the classic mold of idealism-sensualism. We must stand with the other and withstand the other, in order that in creative opposition this shared world may be fully experienced. Man is the living being who ingathers the other in eros. Any experience of the other as bodily is at least minimally "sexual," if sexuality describes the orientation of the whole man toward a fellow human. The handshake, the arm on the shoulder, the embrace of welcome, are all "sexual" gestures, in that they are bodily contacts between humans. Classic asceticism abhorred the sense of touch, and its main tenet was "Do not touch me." "I am my body." Marcel, Merleau-Ponty, and others use this formulation. My body is not an instrument I use nor a prison I inhabit; neither is it something I rid myself of in the fantasy of pure contemplation or in the reality of death. When I am rid of my body I am rid of myself. I am also rid of the other. I end in pure solipsism and in annihilation. Sexuality is the guarantee of my basic humanness.

If primordial experience is but a fancy expression for awakened awareness, then primordial sexuality, of which genital sexuality is a heightened form, is an awakened sexual awareness. Genital sexuality is thus a heightened awareness of the bodily other rather than of the body of the other. It is the heightened awareness which is the source of both the exquisite pleasure and the "pain" of sexual intercourse. But this pleasure and pain is not solipsistic except for the ascetic; for the human being it is shared pleasure. Hence, primordial sexuality, as genital, implies the highest kind of sexual responsibility. But such responsibility is not merely an abstract quality of character; it is a fleshly responsiveness and response. In genital awareness we become aware not merely of pleasure; we are aware of the other in heightened and intimate manner. Response in this case is not verbal as much as rhythmic response. We become aware of the other's subjectivity incarnate as we engage ourselves rhythmically – one might say musically – with the other. And thus man and woman fuse in the ecstasy of love making. The best cure for sexual-eroticism is not pornography but the response of love. In this response anatomy and mechanics are not thematized, as the partners engage in the art of love and are carried away in the floodtides of mutual awareness. Anatomy is for the scientist, the metaphysician, the theologian, the voyeur, the Don Juan.

Genital sexuality is not "animalistic" or merely instinctual except to the metaphysician. It is especially genital play that upsets the ascetic; for, it would overturn his solipsistic world. It would make him share himself with someone. It would turn his heart to flesh. Genital sexuality is not distinct from primordial sexuality but is simply a closer focusing thereof. It can be narrowed down to physical encounter, but basically it is the intensified engagement of primordial sexuality or eros. The sexual impulse is that by which I am naturally impelled toward the other in the opening up of world. It becomes an ecstatic sharing of a world with another. The "ecstasy" is not merely one of intense emotion and sexual pleasure. Rather these result from that *ekstasis* which is a coming-out of self and a reaching-out toward the inmost knowledge of another human being. It is the overcoming of solipsism and results in the meaningful communion of at-oneness.

As a word, sex comes from the Latin *sexus*, a noun formed from the verb *secare*, to cleave. This refers to the cleft or difference between male and female. In mythology (which surfaces also in Plato's *Symposium*) man was originally one but came to be split in twain. Thus the two severed or cleft parts strive to be reunited and regain their original at-oneness. In the Jewish Bible Eve was cleft from Adam, and they were told to increase and multiply by coming together sexually, by ingathering one another. The Greek word for sex was *physis*, which has also been translated as "nature" and as "person." *Phyo* means to grow and it involves the whole cycle of life and death. For it means implantation, birth (thus *natura*), growing, maturing, flowering, bearing fruit, going to seed, dying, and being reborn. This is the *telos*-cycle of life. It *is* sexuality. Life was conceived *as* sexual. Sexuality in our sense was not thematized, since it was regarded as a natural part of human life. Asceticism would have been incomprehensible to the Greeks. It was a creation of neo-Platonists. Christianity introduced sensuality into civilization because of its ascetical models, Nietzsche's No-sayers, the misborn and misguided, who fled bodily existence and created a "higher life" of fantasy and solipsism.

Merleau-Ponty is really closer to Kierkegaard than is Heidegger, even though he does not emphasize anxiety. Merleau-Ponty writes of the bodily dimension and of the "ontological body" as our true subjectivity, rediscovered in temporality and in sexuality. Our body is the "subject" of all that we are, i.e., it *underlies* every experience and every act, everything by virtue of which we *are* human beings in the fullest sense. By bracketing sensual-eroticism we uncover human eros. The intellectualisms of classic asceticism (still prevalent in societal mores and in the methods of science) are in reality the debris of an unsuccessful epoché. Don Juan, the idealist, is in anxious flight from his true body and thus his return to body as such is caricatured. (Science has forgotten its subjective groundings and can view body only as an object, i.e., as basically distorted.) Don Juan *is* anxiety: the idealist caught in a bind between Scylla and Charybdis. Insofar as this is brought out in Mozart's music, Don Juan *is* music. The basis of *Angst* is a musical eros, not just the story of Don Juan set to music, but Don Juan's problem demonstrated in Mozart's score, emerging vividly in such sensuous sections of the opera as the duettino of Scene IX, "La ci darem la mano."[13] In *La phénoménologie de la perception* Merleau-Ponty demonstrates that he was fully aware of the sexual-musical implications of the bodily dimension. But he did not derive this from Kierkegaard; rather, his sources are Husserl and Sartre. Heidegger's *logos* is still too philosophical. The 'ingathering" he writes of needs to be described in more bodily terms to include eros. Man is indeed the *Gatherer*, in that he collects food in the forests and fields, gathers his loved one into his arms and is ingathered into her body, recollects himself in thoughtful meditation (and thus is the *Thinker*), puts sounds together to make music (and thus is the Com-poser), gathers together with other men (and thus forms Society). Don Juan was unable to be ingathered in the sweep of eros. He lost himself in the abstraction of the physical-erotic body. Don Juan was essentially alone.

III. THE BODY-EGO VS. BODILY-SELF

Christianity, Kierkegaard writes, introduced sensuality.[14] Don Juan is the unacknowledged child of the ascetic. The sensual-erotic did not exist in western culture until idealists preached the "higher life" of the soul and philosophers spoke of the higher thought of the "mind." The

[13] *Ibid.*, pp. 111-116.
[14] *Entweder/Oder*, II, p. 51.

negative effect of this transcendentalism was to make man exaggerated-
ly self-conscious, especially of his body, now viewed as physical and
thus as erotic. The higher life of the "spirit" became an anxious flight
from body, and this under the guise of transcendence. The Platonic
dualism, though an interesting historic theory, was proclaimed as
fact and became of the very fabric of Christian culture. And though it
was originally but one of many Christian perspectives, it succeeded in
suppressing all others or subsuming them under its aegis. The flight
from the human body to the soul abstraction produced yet another
abstraction: that of the body, now regarded as physical or sensual and
thus as sexual-erotic. Sexuality as sensual-erotic is but a mentalism.
Moreover, in making the body an object of lust the classic ascetic
depersonalized and dehumanized both himself and others. True
sexuality is that not of the object body but of the body subject, i.e., of
the human person, the human being, the human who *is* eros body. This
love body is no longer merely erotic; and as such it is paradoxically
more desirable. Could Don Juan but have discovered in the other the
true eros body, he would not have had to go on from conquest to
conquest, for he would have been delivered from the need to conquer
and could have given himself up to the encounter of love. He would
have found his eros world and would not have had to go on and on,
seeking for it in vain in the mere quantity of his sexual exploits.

Don Juan is still very much with us in the ordinary man, in that he is
a product of post-Christian culture, a culture which has not yet succeeded in overcoming tradition. For the "normal" man still looks on the
human body as erotic rather than as bodily, and thus his preoccupation with sex as biological. (Thus the materialistic approach of classic
theology toward birth-control and its fixation on techniques.) This
preoccupation of the classic ascetic with sex is something he also projects into others. It belongs to his general religious depersonalization.
For the flight to a religious or personal metaphysics begets a fixation on
physique. Crimes of the flesh are the result of crimes of the spirit. And
at the level of the ridiculous, pornography is the natural backlash of
spiritual fantasies. It is from the exaggerations of soul cult that body
cult, its natural complement, arises. "Spiritual love" claims to be a
sublimation; but it may well be a basic perversion of eros. For, since it
is only ideational love, it is not love in fact. It exhausts itself in beautiful
words about love; but the other is never really encountered. For love is
tested not in the fantasies of the recluse or of the romantic mystic but
rather in the deed of encounter with man, and thus only with man's

gods. Spiritual love seems to be love in bad faith. It is solipsism, grandiose words and good works notwithstanding. The ascetic is intent not on bodily encounter but on "saving his soul." And yet he ends up in solipsism; for, it is not a question of "saving" one's "soul" as much as of making *whole* one's *life*. How can one make one's life whole without love? How can this be a true wholeness unless love become incarnate? How can love exist apart from primordial sexuality, if it is to be truly human love and not some theological fantasy? How can it become truly fleshly love unless one know how to make love? These are questions the ascetic in Don Juan was faced with, questions he was unable to cope with. His solution to all problems was simply to go on to more conquests.

Fixation on the physical easily becomes boring; but the classic figure of Don Juan is always fascinating, for he is tossed between the metaphysical and the physical, i.e., between romantic love and physical eroticism, and never seems able to discover eros itself. Without the alienation that the concept of soul represents there could be no physical body, itself an alienation from the eros body. Yet such is the grounded nature of man (he is of-the-earth as *humanus* and on-the-earth as one who takes his stand there and thus is enabled to stand up and stand out) that either soul or body alienation is impossible to maintain. Either one passes from one to the other or one regards eros merely as a "temptation" of both. For the ego – whether it be soul-ego or body-ego – is inescapably grounded in one's bodily self, of which ego is but an abstraction, i.e., a withdrawal, a facet, a shape, a surface. The abstraction or withdrawal is one of anxious flight from the love body. Now a return to eros body is effected not in the state of psychological anxiety but in *Angst*, i.e., in the opening up of wonder. It is an invitation to the adventure of encounter. Don Juan tries desperately to effect some sort of encounter (his sexual "impulse" literally impels him toward it), but he never succeeds, because his attempts are all carried out in the modality of the sensual-erotic, and thus he is perpetually doomed to failure and to trying all over again endlessly. He never really encounters; he only meets, even when in erotic embrace. Physical body meets physical body, but in the erotic intertwining of both no intercorporeity results,[15] because subject does not encounter subject;

---

[15] E. Paci, "Per una fenomenologia dell'eros" in *Nuovi Argomenti*, nos. 51-52, 1961, p. 74, "L'incontro tra i genitori, il loro reciproco compenetrarsi, la realizzazione (ancora *l'Erfüllung*) del loro accoppiamento, della loro *Paarung*, del loro reciproco sentirsi nelle *Einfühlungen* intercorporee ('intersoggettività e qui intercorporeità'), tutto questo, dal passato non presentificabile si rovescia nell'avvenire, nell'attesa, nel desiderio di un avvenire procreativo."

and thus there is nothing that *underlies* the experience. The experience is no experience, since it does not root in eros. It is mere rutting. The partners are not really doing what they do; they are mere performers, mere play actors. Essentially, they are observers, not participators. They observe one another's bodies and their own pleasure. But there is no bond between them, no true *inter-est*, in that there is "nothing between them." In his very erotic proximity Don Juan is not close to but distant from Zerlina; whereas eros begets nearness even in distance. Here we see the futility and essential deception of both Platonic love and its counterpart, sensual-erotic love. Both are depersonalizing. No encounter is possible under such abstract circumstances; for, as an abstraction the ego is withdrawn from the other and drawn in upon itself. It cannot be truly outgoing. It must always draw in; it must ever implicate the other in its amorous ideology. It can only "have" the other as an object satellite; it cannot truly "have and hold." (Thus the other can actually look "had" after such a meeting. But she cannot radiate the joy of having been held.) The ego cannot encounter the other in eros movement, because eros is not ego-centric. It is concentric: two *are* in one flesh in the intertwining of *human* bodies, as together they are gathered into the center of eros.

But in the ingathering of eros the "two in one flesh" do not engage in some solemn theological or ontological enterprise. Sex has been too solemnized. In the act of love there is a whole range of emotions, from the trite to the ridiculous to the sublime. For a mature sexual bond a sense of humor is required. For the act of love is not only intense pleasure and exotic emotion, it is often simply fun. Where there is no wit and humor the act of love could become a bore: either too physical or too spiritual. Perhaps a sense of humor is the phenomenological moment which maintains eros as such. *Kama Sutra* gymnastics, which apparently succeeded well enough for classic Hindus and Indians, seem excessively complicated to Christians and post-Christian man. It is especially here that a sense of the ridiculous is in place. It is obvious that in authentic love and laughter neurotic preoccupation with eroticism is banned.

"I love you" is an important but ambiguous phrase. The concept of love in our culture is ordinarily a result of romanticism. What we do not recall is that romantic love started out as court love, i.e., as a love of courtesans. It was not to be found in marital love, where feudal duty and offspring were emphasized. And so typical Christian husbands did their duty at home and found "love" outside of it. Christian wives found little love at home and a good deal of duty in the form of house-hold work as well as of sexual "debt." They therefore turned to cooking and to gossip. Happy the man whose wife's preoccupation was not gossip. The word, love, could be redeemed, however, if we were to cleanse it of its historical Platonic as well as extra-marital connotations and spiritual obfuscations. If spoken at the height of the act of love or after it, "I love you" best expresses what has already been "said" in bodily manner. The words are really unnecessary, unless they are meant to replace something missing. If their bodies have already expressed love, the verbal phrase is a kind of embellishment. If Don Juan's body tells his partner he really does not care except for himself, then the words are at best meaningless and at worst a lie, whatever their status as a proposition.

The soul-ego turns out to be nothing but a shadow-body. And body-ego is itself but an after-image of the truly human body as bodily self. The flight from body begets a return to body but now as caricatured, i.e., as merely sensual. Departure from bodily self-awareness begets self-consciousness: we become spuriously conscious of self as a body. Thus the false shame or modesty of the disrobed Platonist, who is overly conscious of his nakedness in the presence of the other. Returning to true body entails overcoming also the self-consciousness resultant upon alienation. For such self-consciousness is not awareness of self as such, but rather attention to a specious self, the body ego. The person who makes an ontological return, i.e., returns to what he already *is* naturally, rather than merely diverting his attention from soul to body or vice versa, delivers himself from self-consciousness. He stands beautiful and strong "before his Maker," like Adam before his "fall." Pseudo-sexual attraction of body to body is the result of the false spiritualization of body. The eros body, best portrayed by classic Greek sculptors, is a work of art, of power, of magnificence, because it is a body of proportion and beauty that does not allow of reduction to the merely physical and needs no spiritualization. In post-Christian times the reduction must be worked the other way. Body-ego must be reduced that the true love body may emerge. Soul-ego must be banished to its solipsistic otherworld, lest it continue to breed its antibodily virus, infecting man with its strange ideals and deceiving him that the human body is merely erotic.

IV. ALIENATION: THE EGO AGAINST THE SELF

Both Faust and Don Juan are the personification of things begotten by a society alienated from a true bodily self. Faust personifies evil, and Don Juan the sensual-erotic. Faust is the unacknowledged offspring of the righteous man; Don Juan is the rejected child of the ascetical monk. Both are the result of dualisms introduced by classic Christianity: the dualism of good and evil, the dualism of spirit and body. The attempt to become good entails the covering up of inherent evil; the attempt to become spiritual brings with it the compensation of sensualism. With the best of intentions and taboos the one abstraction leads to the other. The "pure" soul is the parent of the "impure" body. And both are meaningless to him who achieves a return to ontological body. The ontological body is not an object in an area of cleanness cordoned off from the unclean. It is not light that parts with darkness. It is surely not the area of the dark and obscure, as Descartes would have us believe. However, it may well be the area of the unknown to us in our present alienation. To the ascetic it may appear as the area of temptation and even of unbelief. But such idealism and its concomitant eroticism are problems only for cultures infected with Christian Platonism. This is not to say that the classic dualism did not produce certain treasures for our culture. The body of philosophy, music theory, and mathematics, attests to a certain greatness of the disembodied mind. But even Descartes was dissatisfied with it and felt that a new mathesis was called for. In our day the rediscovery of ethos and of eros places ethics and erotics in proper historical perspective. The uncovering of the ontological body is necessary for a proper appreciation of the historic place of classic philosophies. The discovery that music is the expression of a fundamental eros helps shed light on the proper place of any music theory past or present. In this especially Kierkegaard's Don Juan essay is of inestimable import.

Negatively, metaphysics has been a form of alienation, whether it appears in art or in the life of such a figure as Don Juan. Soul and body ego work against a fundamental *being*. When I say, "I am," I abstract the *I* from the *am*, even though subsequently I put them together in a statement. Grammatically the "I" is substantive and the "am" is a verb. For us the "I" is the most important word in the statement. "*I* think, and thus *I* am." We have detached the ego from the existential activity of the word, to be, which henceforward is regarded as unnecessary, or as a mere link word with something else which I am, as e.g.,

"I am a musician." The verb, to be, is thus rendered innocuous and is reduced to an abstraction. It means nothing; it only links me with my musicianship. Being equals nothingness. It no longer has anything to do with growth, fruition, living and dwelling on the earth or with that happening we know as eros. The word, like present day currency, has undergone drastic inflation. It has become valueless and meaningless. And only the disparaged metaphysician still uses it in philosophy. The grammar of our modern languages, the child of historic logic and metaphysics, forces us to separate and categorize ego and being. Not so in classical languages, such as Latin and Greek, where *ego* was used only in certain cases for emphasis, where *sum* and εἰμί meant "I am" not just "am." In these languages the "I" was ordinarily not used; ego and being were one. But in our post-logical grammar ego is set against being, indicative of what a state our civilization has been in. Being has been degraded in favor of ego. Yet ego is but an abstraction from the self, which is at one with bodily being. Being is not metaphysical, unless we simply accept the alienation that history foists on us. Being is *physis*: eros body, that which guarantees our remaining human as "human beings." (Apparently, the Greek *phy-* and our word *be* are cognate.)

When we add to the "I am" and thus say, "I am a musician or a philosopher," the "I" and the "philosopher" or "musician" are regarded as more important than the "am." We could omit the word and simply write: I = musician, philosopher. Yet the link word, am, is a bond which does not merely bind together two words in a proposition but binds me to my métier. My métier is not simply a job I do; it is my whole way of life. In it "I find myself," i.e., I discover who I really am: I discover or uncover my *self*. This is especially true when I say, "I am my body." This statement is necessary to counteract "I am a philosopher." The "am" holds the full self and with it the world and the other. I cannot "prove" this, but I can probe or test it, i.e., I can discover experientially, though not experimentally, if I live in a bodily manner (and love in a bodily manner) or simply exist egotistically. The "am" of "I am" is the ontological ground from which ego is derivable in any form, even as alienated. It *is* my ontological body, i.e., my body as non-metaphysical and thus also as non-physical. And when I actually speak the sentence, "I am my body," it is my ontological body, i.e., my body as it *is*, which forms these words and says them. The statement is not a dry proposition.

Yet ego attempts to abstract from all this and tries to become a ground unto itself, apart from truer self, world, earth, and the other. Egoism is the self-grounding of ego within itself as withdrawn from eros body which the self *is*. This self is not alone by itself but is one with the world and with the earth. The human being is the earth-being who comes from the earth (*humus*) and will one day return to it. In rejoining our self we return to the world and to others. When I separate the "I" from the "am," I isolate and alienate myself from self and from the other. At most I establish an artificial relationship with the person "out there." But the road back necessitates doing more than establishing a relationship; it entails overcoming the artificial split between ego and self, and thus between self and the world as such. The world *as such* is not merely my own individual world, however I define it, nor is it merely the actual cosmos our orb knows of and into which our cosmonauts launch themselves. World as such is discoverable in the uncovering of my self; there it is experienced as the world in which the other is already there. Whether this world can be exactly formulated in words is a moot question. At best it can be hinted at in description. It is best experienced rather than verbalized. The ego cannot understand this world, since such a world eludes "reason" and its categories. The ego in fact refuses to discover this world – and this out of anxiety for fear of loss of control or loss of identity. In the face of true self the ego is in a state of anxiety. Since ego is anti-erotic, it stands in *Angst* of eros.

"I, the philosopher" is the full substantive we seek. The "am" merely does service as a link between the two words. Strange that an abstraction links the thinker with his ego! Yet the ego can be viewed simply as an excrescence or outcropping of the self. (The ontological body is the dissolution of the ego; the body-ego is simple dissoluteness, as in the case of Don Juan.) How is the ego an extension of the self (like the pseudopod of some amoeboid creature?), and how does it become an "excrescence"? When I say I am a musician or a philosopher, I indicate that, if I am serious about it, I have thrown myself into the profession. Insofar as we throw ourselves completely into something we grow in it and fuse our being with it. But this casting of oneself in the role of musician or philosopher is in reality the development of an important facet of one's self. The self extends itself as musical consciousness, as philosophical "ego." This ego finds itself to be a part of a collective of other such egos, and throwing oneself into a thing as collective can entail submerging one's self in the community as such. This is a critical point, since one can find either fulfillment or ruin in a collective *as* institutionalized. When the collective is political or religious (or both) one risks debasing the self in favor of a collective ideological ego. When this poses as sublimation, one is on dangerous ground. When the ego, as extension of the self, finds itself caught up in abstractions, it is in danger of losing its moorings. Loosened or severed from its bodily grounding it becomes a free-floating outgrowth or excrescence. Thus the ascetic escapes to his ideal world, be it that of pure forms or of scientific objects, and the reality of the bodily dimension is lost on him. He becomes "alienated," i.e., ontological reality becomes strange to him; he is estranged from his own body. He becomes the Outsider to his own life and to eros.

The eksistential ego stands-out from the bodily self, not as an ideological excrescence or as a case of psychological alienation, but as a given stance of the self. The self stands erect as ego, in that I "shape up" and stand up as the musician or the philosopher I *am*. With regard to eros, one's bodily self becomes doubly erect in the presence of the loved one and in the act of love, most often accomplished in a prone position, as one lies against the earth; the eros-ego stands out as the rhythmic link between man and woman, not unlike Antaeus who regained his strength (rather than was "depleted") in falling back into the embrace of Gaia. The eksistential ego stands out (ek-sists) into the other, and in opening up world as such discovers earth as such. But it does not extend itself into some metaphysical Beyond. Thus is discovered the meaningful "transcendence" of the ontological body as inwardness, i.e., as it were, as "in-scendence" or immanence. Such inwardness is not introversion or inversion but a living *in* the world and a dwelling *on* the earth. The "I am" yields to the "We are" in such a way that the union of two in love yields the fruit of offspring.

The difficulty with ego in its ordinary meaning is that it is regarded as ex-centric and centrifugal, i.e., it pulls away from and spins itself out of the force field of the self. The philosopher throws himself into his work and becomes a professional reasoner or even a functionary rather than one who thinks. Often the political and economic underpinnings of his position prevent him from professing any "truth," but rather require that he profess an ideology, whether political, religious, or scientific. In recent history intellectuals capitulated wholesale to various ideologies, in order not to lose their positions. And thus intellect was rendered innocuous, if it was not actually pressed into the service of a party ideology. The thinker who is at one with himself is a dangerous man indeed. He is always "in trouble" in an ideological society, for ideology is foreign to *eidos*. Ideologies are built on a metaphysical idea about reality, not on reality itself as it gives itself to the self, as it shapes and forms the self, as it gives us the bodily "idea" of our self. The unexistential ego forms an abstract idea of itself rather than allows the true self image to emerge. Whereas the existential ego is a manifestation of the bodily self and is in true polarity with the self, of which it is a special structuring, the unexistential ego is split off from the bodily self and becomes discontinuous with it. Perhaps this is the root of that withdrawal we call schizoid. The ego wants no true contact with the self, particularly with the eros body. Yet it is ever forced to return to body, in however a psychotic manner.

The estranged ego in its neurotic return to body regards body as mere body, as physical, as material, as sensual-erotic. The body is regarded as sexual in a narrow biological sense. The sensuality of the human body and the sensuality of human art, such as music, was the result of ascetical Christianity. Thus Kierkegaard. His profound admiration for Mozart's *Don Giovanni* stems from the fact that in this opera he found sensuality at its height both in the person of Don Juan and in the music itself. Music is thus "demonic" in its import for Christianity, which claims to be the religion of the spirit as such. In language, which has metaphysical pretensions in regard to meaning, the sensual element is regarded as subordinate or even as unimportant. But music, even though it has been conceived along metaphysical lines, has nothing to do with cognitive meaning and is expressed in the sheer sensuousness of sound as such. The early Church Fathers recognized this and thus banned the sensual sounds of reed instruments from the church. The Reformers recognized this and some even banished the organ as the "devil's bagpipes.' In *Don Giovanni* the demonic element of the human person and of music as a human art are combined. The ascetic must fly from it as from a temptation of the devil. For things bodily, whether a man like Don Juan or the music of a man like Mozart, must be regarded as erotic and thus as a threat to the classic ideology of the ascetical egotist.

The ascetical flight from body is a departure from bodily *askesis*. The ontological *askesis* of the body entails stripping off of what is inauthentic in us, like egotism of the body or mere eroticism for its own sake. The classic ascetical ego not only prevents true *askesis* but creates eroticism as such. The flight in anxiety from true body preconditions the return to body, either a real return or a fantasy return. The ascetic (and by this is meant by no means only the classic recluse) chooses sexual fantasies as his mode of return rather than the eros body. But, as the case of Don Juan easily demonstrates, it is precisely these fantasies that block true eros.

The "higher self" of idealism is not the true self at all but only its pale image. The higher ego and the higher life (whether of classic monasticism or of modern science) is a prideful self, aloof from the exigencies of being truly human, insulated against the challenges of an earthly existence, i.e., of one as fully human. The gradual alienation of ego from self – to the point that today we may not even speak of the "self" except in terms of a mystic postulate – can be traced in the history of our culture. It is concomitant with the thematization of "consciousness" as ego-awareness, rather than as awareness of bodily self. Early in our culture the body was disparaged, the senses distrusted, the soul and the mind glamorized. The rise of ego-consciousness and its house of intellect goes hand in hand with the suppression of ontological awareness. It is not just a case of the "forgetfulness of Being"; it is a question of the suppressive forgetfulness of Body. One did not just happen to forget bodily Being; it had to be disparaged and submerged under the crystal palace of the mind. Ego-consciousness is the result of onto-logic.

In practice the alienation of ego from self takes place in the learning of a language as a grammatical system rather than as a complex of sound-symbols. Meaning rather than intonation and tonality is stressed. Grammar was the child of onto-logic, itself an alienation from *logos*. A small child learning a language is not merely interested in the labels we put on things, as Augustine would have us believe, or in language games (Wittgenstein). The child is interested primarily in joining our world, and the port of entry is the complexity of sounds he hears us using. The labels for things make sense only in terms of the societal world he wants so desperately to join. The sounds could be any sounds; it is indifferent to him. But they happen to be English sounds. He will take us at our word, whatever it is, English, Russian, French. We could make up words and he would accept them, he trusts us so completely. The important thing at first is not the "meaning" but the sound pattern, and anyone with children has heard the child practicing the sound patterns as such, as he or she learns the language. Words he has mastered are then placed judiciously in the sound patterns. The child learns to "fake it" before he ever comes forth with a correct proposition. This "faking it" is his attempt to join us in our world of sound. Specific sounds with object meanings are significant only as a part of this family world into which he was born. Words come through to the non-metaphysical mind of a baby not as carriers of "meaning", but as tonalities that establish a link between him and the world of the family as such.

In the learning of grammar, which comes after the learning of the language itself, the accent is taken off the bodily warmth of words and the abstract structure is stressed. The structure is quite ascetical. The grammatical word has its basis not so much in the warm sound of utterance (and the warmth is generated by eros!) but in the logical articulation of words as written and as thought. The impositions of classic logic on thought and language is a matter of twelfth and thirteenth century history of pilosophy. But apparently it goes back before this, for the classic poetry of Anglo-Saxon and a good bit of its prose shows that Latin grammatical and literary forms were imposed on the progenitor of English. Later French and other overlays laid Old English waste. The wonderfully live world of Anglo-Saxon literature came to an end, but many of its strong four letter-words (like the word, love) survive today. Another such word is *eard*, our earth (also *eorthe*). From such stems one could form a verb, *eardian*, to dwell. Today we merely inhabit the earth; and in inhabiting it we besoil it. Our habitation of the earth is mostly above the earth; we despise an earth dwelling, though we seem to go along with the present fad of earthenware in our sterile homes. The grammatical word is fixated on the letter (*gramma*) and its "proper" copulation with other such letters. The word, to be, has little meaning of its own, Webster informs us, but it serves as a copula. This is a hysterically funny statement. Being is "only" copulative! In grammar even copulation is neutralized. The metaphysician spooks everywhere in the haunted manse of classic grammar. The cruel countenance of the ascetic wrought its vengeance on grammatical forms long before we were born. But somehow language survives as sheer sound, sound that introduces us into the world of our parents and family. Even the twelfth century logicians had respect for the *vox*, the human voice as *sounding* the words. Perhaps something in them recalled how they had actually learned to speak: from shaping the sounds they heard in the family circle, in the warmth of being a member of an eros society. And it was in this learning to speak, i.e., in this joining the sound world of the family, that they began to "think" and structure their thoughts "logically."

The written word, and its after-image, the mental word, are inadequate substitutes for the living "word" which the whole man already *is*. Even the verbal word, as lingual, is a narrowing down of the bodily word. We must begin at the beginning. Man, as a bodily being makes use of sound and thus comes to speak so as to be part of a world. But he speaks with his whole body in movement and gesture, as is still observable among Latins. Within this complex of movements, all of which are oriented toward a world of human subjects, his tongue is also engaged in making specific sounds, which detail the movement and the joining of a world. This stage is important, given that it is good to "make the right noises" and "choose the right words" if we want to be accepted in a new world. But even here body language is crucial. We may succeed in saying the right words, but the intonation and the body movement must also be right. In speaking to a purple dragon our intonations must imply at least modesty, if not that abject humility evidenced in the prefaces written by Baroque music theorists, as they dedicated their work to the prince patrons of the time. Moreover our bodily pose must also imply a certain meekness. We must not stand too tall, for fear of arousing the anxieties of the prince. Only in rare cases was there a prince secure enough to admire men who stood tall and erect. In addition to all this we must adopt the acceptable curial language with all its embellishments. It goes without saying that the grammar is correct. But all of this is far removed from the naive world of the child, who could cry out that the emperor had no clothes.

The last thing we look at in the human context is the grammar of speech. Apparently this is also true from the historical perspective. However, this is not to imply that grammar is unimportant. It is just that a prince can forgive a grammatical slip; but he can never forgive improper body language, which is a threat to his own exalted position. The prince cannot allow anyone to be more erect than he. (For he is the chief of the harem as well as lord of the land.) We arrive at *gramma*-language only as a detached study of what has already *happened* in the concrete world. When such language becomes mental and evolves into a recreational dialectical skill, we are in the area of play, as is evident from the twelfth century in the works and pomps of Abelard. But here we are justified in asking whether or not grammatical language, for all its usefulness in dialectics, is not several times removed from the full existential "word" that man, as human, is.

Our culture introduces the school child to the historic fact of grammatical language without imparting the perspective that would correct emergent illusions about it. The child is taught to separate words and consider them out of the context of the live world. Word and deed are separated, as we inactivate words and detach them from the world of sound in which they were born and in which we first experienced them. It becomes possible to think and even say many things without having to do anything about it. A schizoid world results, one in which the political animal is carefully segregated from the grammarian and he in turn from the poet. Thinking and saying are abstracted from the reality of living and doing; and thus it is enough for us simply to think or say we are in the human world. We do not have to "do" it in the sense of Merleau-Ponty's *"faire le monde."* With regard to the eros world this becomes more intelligible in English, for we not only speak about or write poems about love. We *make* love, and thus join the world of the beloved in an intimate manner. In making love rather than in just talking about it we bridge the distance that separates, and reaffirm the bodily bond between two human beings. The language of love is primarily body language.

Heidegger tells us that "saying" is a "showing." Wittgenstein seems to imply the same thing, for when we say anything whatsoever, the world "shows itself" in our speech. In the eyes of a Wittgensteinian Heidegger's mistake is his attempt to give verbal form to the "form of the world," which indeed shows or manifests itself in the saying of any statement or complex thereof, but which cannot be properly elucidated therein, particularly not by means of a break with everyday language.[16] The world can show of itself but it cannot be articulated in everyday language. Oddly, Heidegger seems to agree with this, for he breaks with ordinary language and attempts to sound out the possibilities of speech before and beyond everyday usage, though he also sounds out the content of idiomatic phrases and words. Wittgenstein's "world" is a kind of Mona Lisa, who beckons with a "smile" but does not open her lips to speak. Heidegger tries to take her by force, as it were. And thus the incredibly fascinating stage of *SZ*. Yet in all this Heidegger, too, tells us that Being both reveals and conceals itself. It reveals of itself in classic propositions and scientific statements but it also hides therein. The world is both revealed and concealed in Heidegger's thinking. What effectively obscures this world is not so much the Quixotic language of the early Heidegger as the *monde-en-idée* itself, which makes such linguistic gymnastics necessary in the struggle of

---

[16] K.-O. Apel, "Wittgenstein und Heidegger. Die Frage nach dem Sinn von Sein und der Sinnlosigkeitsverdacht gegen alle Metaphysik" in O. Pöggeler, *Heidegger* (Köln 1969), pp. 358-396. Cf. also M. Dufrenne, "Wittgenstein et Husserl," in *Jalons* (Nijhoff, 1966), p. 188 f.

man to emerge from under historic overlays. Wittgenstein took a less frontal tack than Heidegger and refused to attempt to make Mona Lisa speak. He lets the world glimmer through speech but he never thematizes it as does phenomenology.[17] Perhaps this is being more subtle; perhaps it is less courageous than Husserl and Heidegger. Whatever the case may be, it is not yet quite clear whether or not the subject-world will actually allow itself to be articulated in words we can understand, or whether it will simply "show" itself in whatever saying or in whatever sentence that we say. Husserl designated the life-world as an "anonymous realm" which had hitherto been un-named.[18] Perhaps it would have been better to leave it alone, except that in naming it he guarantees our not succumbing to ontic worlds that emerge from it, like the world of language. Apparently his purpose in naming it was not merely to control it by means of the label, life-world. Whether such a word formulation can then be further designat-ed as "lived world" is a moot question, as moot as the possibility of a truly existential philosophy as such.

According to Heidegger, when we say something we show who we *are*. Obviously there is some truth to that even at the level of everyday language. To say something therefore means not just to come out with a grammatical or logical proposition but rather to manifest Being, i.e., the ontological self as opposed to the ontic ego. The "ontological" self dwells in the world (and on the earth); the ontic ego merely inhabits a restricted field and postulates a specious *monde-en-idée*. The language of the subject rooted in world finds public language a block to full or even adequate expression. He must communicate with others at this flat level of speech; he communes with the other in body language. At times "ontic" languages (i.e., languages "the way they *are*") fall silent. The language of silence is more than the dramatic pause of classic rhetoric or of Baroque music. It is the suspension of the control of a narrow range of speech, in order that the full spectrum, which in-cludes body language, may emerge. This is not the destruction of grammatical language; it merely lends it perspective.

Western education introduces a child not only to language in the abstract, i.e., as it were, to "ascetical" language disengaged from the warmth of live words as erotic sound (e.g., the "sexy" voice); but to the historically and politically conditioned usage of a given society or segment thereof. Thus words take on a given "public" meaning,

---

[17] *Ibid.*, p. 365.
[18] E. Husserl, *Die Krisis der europäischen Wissenschaften.*

eventually regarded as "standard." And since people stand according
to a standard and fall without it, or at least seem to fall in the eyes of
the public, it is necessary to speak the language and observe the norms
of the particular society, in order to gain and retain status within it.
Thus Don Juan, cannot simply declare his full intentions to Zerlina;
he must coax her by taking her hand, leading her to his house, and
telling her that it is all meant only "to alleviate the pangs of innocent
love."[19] The gestural and linguistic ritual is one dictated by society. In
this way their "sin" is socially acceptable; to seduce without this ritual
would have been a socially unpardonable crime. And, of course, the
"meaning" of the words of seduction is greatly enhanced by the sensuous quality of Mozart's music.

Heidegger, a great admirer of Mozart's music,[20] maintains in SZ
that Being stands in opposition to the standards of static society. The
language of everyday and of everyman is the language which "one"
speaks rather than what I say, or really mean to say. One speaks English as one speaks it. Thus ordinary language is not my language or
yours but the language that an impersonal "one" learns to speak.
Public language submerges us in the inauthentic self-of-the-One. This
societal One is nobody in particular and everyone in general. We
speak no-man's language; and this is the source of our "clear and
distinct" concepts. Viewed in such a manner, the situation seems
somewhat Kafkaesque. The world of the subject cannot be articulated
in the language of the outside world. In this Heidegger and Wittgenstein seem to agree; but the conclusions each draws with regard to
philosophical method are literally worlds apart. Heidegger hacks his
way through the forest of words. Wittgenstein skirts the problem

---

[19] *NMA*, II, 5, 17, pp. 114-116, "Andiam, andiam', mio bene, a ristorar le pene d'un
innocente amor."

[20] M. Heidegger, *Der Satz vom Grund*, pp. 117-118. Here Heidegger breaks into his lec-
tures (*Neunte Stunde*) on the principle of sufficient reason, in order to commemorate the two
hundreth birthday celebration of W. A. Mozart. He quotes Mozart's letter on how he receives his musical inspirations: in audial imagination ,the entire composition presenting itself to him at once. In commenting on this Heidegger makes an interesting aside concerning
the unity of creative intuition and audial perception or imagination: "Die verborgene Ein-
heit dieses Er-blickens und Er-hörens bestimmt das Wesen des Denkens, das uns Menschen,
die wir die denkenden Wesen sind, zugetraut ist. "This is an interesting insight into what
the present author has called "synesthesis" in other publications (cf. "Toward a Phenomenology of Musical Aesthetics" in *Aisthesis and Aesthetics*, eds. Straus and Griffith (Pittsburgh,
1970). In *Der Satz vom Grund* Mozart is called "der Hörendsten einer unter den Hörenden
gewesen. . ." This does not mean that Mozart's hearing was merely acute; for Heidegger
had already made the point that musical sound and modulation can be employed to illustrate the ontological leap. Philosophy "in a new key"! Heidegger implicitly ontologizes
Mozart's music or at least initiates the process. Cf. also his contribution on Mozart to *Le
Grand Larousse*.

entirely, so it seems, but he knows it is there, and he makes use of timber gathered from the forest.

The societal ego, which feeds on suppressed eros, is best seen in the political or religious collective. Here the personal ego fuses with the social ego and one attains "communal consciousness" as a "good fascist" or as a "good Christian." True personality seems submerged in the collective personality. And in fact the individual person is allowed to exist – often to eke out his existence – only insofar as he is a "true believer," i.e., he sacrifices himself to the ideology of the community. Being a true believer often means losing faith in self, as in Stalinist Russia or Nazi Germany. The truth the true believer professes to believe in is untruth, the nemesis of his truer self, faithlessness as concerns primordial belief. The collective is caught in its own ego; it is self-sufficient, closed in on itself, closed off from the other: *en soi* (Sartre), demonic (Kierkegaard). The political collective makes the authentic city (*polis*) impossible. In such a city the human being finds himself without a city (*apolis*), and thus apolitical. The false collective puts men in bondage and tries to dissolve the natural bond between man and "nature." It tries to manipulate language in order to control man. Thus the rise of 'official" language, of curial language, of ideological language, whether political or religious.

The ontic collective with its language and thought controls tries to prevent the ontological city from forming, i.e., the city as it *could* be: a place to dwell, a truly existential crossroads. The collective sets up the "ontic" city, i.e., the city as it *is* (conditioned, controlled, manipulated) as a model; it derides the very notion of an "ontological" city, i.e., the city as it could *become*. The existential city cannot ek-sist; it cannot stand-out amidst the debris of the degraded and debauched city. It must go underground. The city, as it *is*, employing language as it *is*, becomes not a true city but a borough or burg. A burg is a fortified town, closed about by a wall which keeps everyone in and everything else out. And the people in this borough or burg become good burghers. They are the basis of the typical bourgeois society. The bourgeois collective systematically displaces the true self, substituting selfish contentment, euphoria, ideology in its stead. Our whole education in such a society has been a systematic seduction. We have been wooed and won not by the genuine self but by professional ego. Name, fame, honor, position, and economic rewards are all based on ego. And so is the bourgeois morality it engenders. Morality becomes simply the justification of what the society is doing, whether it be killing helpless peasants in Indochina, teaching the young false ideals, or simply making money, in order to attain some measure of brittle power. All this at the expense of eros. Pornography is hypocritically banned as a threat to the morals of the country; but no one worries about killing human beings. War propaganda is based on "body count," the dead bodies that were capable of love and further life, but were sacrificed to an ideology. The phrase, "body count," is officialese; it is meant to show that quantities of bodies equal victory of some sort. The body is here degraded to being just an item on some Pentagon chart and a factor in official government deception. In the history of language this is but one evident example of the purposeful manipulation of language. If we rejoin that bodies do indeed count but not as propaganda items, we may be accused of being either poets or radicals. But all we have tried to do in reality is to restore language to itself and take it out of the hands of officialdom.

The "good society" in which the burghers are well fed and content becomes the Good as such. In olden times this was sometimes openly identified with God, and traditional theism became the religion of capitalists, who used it to aestheticize their mode of existence. They indeed led an aesthetical rather than ethical existence (Kierkegaard). This means that the aesthete regarded his true self as something peripheral, Kierkegaard writes.[21] This was Don Juan's problem, also, not just that of the contented burgher; for, the Don's life and his truer self had to yield center stage to his erotic exploits. In our modern civilization the blasé burgher is often involved in erotic exploits, but without the art and the charm or interest of such a colorful figure as Don Juan. One of the reasons why classic communism was atheistic was due to the bourgeois theism of the common burgher. One of the reasons why there are few "red light districts" in Eastern Europe, is that this is justly regarded as capitalistic decadence. The burgher and the Don are the children of western theistic culture.

The morality of the ego, whether personal or social, is alienated from the ethos of the truer self and from eros as such. Don Juan is not immoral; he is amoral. Unable to live the ascetical ideals of Platonic Christianity, but incapable of finding himself and thus his true ethos, he can have nothing to do with morality in a human rather than ideological sense. He simply contributes to decadent social mores. Ethics is the metaphysics or logic of ethos, as eroticism is the distortion

---

[21] S. Kierkegaard, "Das Gleichgewicht des Ästhetischen und des Ethischen in der Ausarbeitung der Persönlichkeit" in *Entweder/Oder*, IV, esp. pp. 321, 325 (Don Juan).

of eros. Whether meta-ethics gets at ethos is a question that goes along with the formulation of a possible meta-erotics. In musicology we speak of the ethos of Greek or of Renaissance music, the latter as opposed to the pathos of Baroque music. We might be able to speak of the ethos of Mozart's music, in that classic form looms so large in his works. Yet Mozart does not disparage the emotions and in his own way he partook of the "sensitive style" of the pre-classical composers as well as of a kind of pre-Romanticism which set the stage for the nineteenth century. The music of *Don Giovanni* has its own classical ethos; but the Don himself is not even ethical in any given sense. It is obvious that the musicological usage has little to do with "ethical" considerations, but it does point to a wider and more interesting use of the word. The musical usage is close to the phenomenological; it is not just the attempt to restore such words as *eidos*, *logos*, *ethos*, etc. It might be regarded as a loan word from musicology and the arts.[22]

The speculative ethic has contributed to our culture's alienation from true ethos. The ethos is not just classical form, as in the musical sense, it is the structure of our character, and the basis of this structuring process must be the eros body. Indeed, for the Greeks $\tilde{\eta}\theta o\varsigma$ meant the position of man's body, especially when he set himself down. It thus became his "seat" or today we might say his "moral position." Ethos is of the body; ethics is the ego's morality.[23] Emerging from ethics into a meta-ethics or into ethos does not imply becoming "immoral," unless we have identified morals with a given historical period. Yet in a society where classic ethics still prevails, though now mostly in the form of "hang-ups," where rewards and punishments (not excluding those of the penal system) favor "ethical behavior," the emergence of a new ethic, i.e., of genuine ethos, is made difficult. Ego rather than ontological ethos, rooted in the eros body, is the power-line along which "thinking" and "action" follows or is omitted. Man is educated to get away from or even to destroy his true self. And this is his fallen state. Instead of standing up and standing out amidst his fellows, he falls away into the apo-stasy of the collective ego. The alienated ego is apostatic rather than eksistent. But this is largely the story, rather than the official history, of our civilization.

Our education may be justly regarded as a gradual abduction from the realities of being in the world and dwelling convincingly on the

---

[22] Since the Middle Ages *moralitas* has had to do not with morals but with musical ethos, even when the word, ethical, is used, as in Aribo's *De Musica* ("Ethica est musica. . .)".

[23] F. J. Smith, "Two Heideggerian Analyses" in *The Southern Journal of Philosophy*, winter, 1970.

earth. The language education makes use of demonstrates this. Educationalese, as it is called, is surely a worse and more vacuous misuse of language than officialese. Both are metaphysical in their approach to language; but they are political as well. We are led away from our true selves and are led to believe that ethos, honor, and learning lie elsewhere: in the façades of our decadent culture. The alienated ego is even idealized and presented as the ideal self, whereas it is not truly ideal but only ideological. And yet it becomes the "accustomed place" (ἦθος) of many people. The customs of peoples become decadent in the sense that they have fallen away from ethos as such. The authentic self is demonized, regarded as evil, as precarious, as dangerous, even as carnal. The latter is the case as regards the eros body especially. In the eyes of the moralist it can only be viewed as erotic. In this sense, despite his immorality Don Juan is a classic moralist. Especially at the time of the awakening of sexual powers moralism does its most pernicious work. Thus the "idealism" of youth, which leads them away from the discovery of self and into the nebulosities of the ego culture. Instead of becoming more truly a human *being*, i.e., more "ontological," the young person becomes more metaphysical. Thus the decadence of western culture, which with the end of metaphysics, is so near to collapse. We must learn all over again to "let go." We must let go mostly of the inauthentic ego with all its neurotic striving. Letting go is the capacity both to let go of ego and to let the eros body *be*. Don Juan is incapable of this, and thus he must spend the remainder of his dissolute days striving for infinite erotic love, as his father, the ascetic, strove for infinite divine love. He cannot let go; he cannot let be. This is the enigma and his tragedy.

V. SEXUALITY AND THE ONTOLOGICAL BODY

Sensual-eroticism is the source of anxiety for the classic Christian personified in the figure of Don Juan. Sexuality is viewed as a threat to the spirit. Yet the ontological body as such does not threaten; it is rather the source of wonder. The path of return from eroticism to eros may be fraught with anxieties, but the wonder of love beckons. The eros of the ontological body, the quest of which is initiated in him who has *Angst* with regard to the physical body, is not merely erotic or sexual. From the standpoint of the everyday man raised in the everyday culture of tradition sex is something special, the object of fears, guilt, and anxieties, rather than what it is to the "natural" man: the unselfconscious subject of spontaneous joy and subsequent maturation. Sex is identified more or less either with biology or with the sensual erotic dimension, both "physical." Unlike Plato the modern Platonist must equate eros with the erotic and bodily encounter must be characterized as sexual-erotic. The ascetic either refuses the contamination of touch or he is anxious about it, even in the act of touching. Even in marriage an area of control must be maintained. He refuses the full encounter of self with self and the aborning of eros world. He forces himself to maintain his ontic ego at all times and thus his encounter with his partner is never ontological; she is always a sex object never the subject of love. Don Juan, the offspring of the ascetical idealist and *as such* the devoted sensualist, never attains to the full bodily dimension despite his many conquests. Since he forever fails, he must forever repeat. His father, the ascetic idealist (whether that be a monk or a scientist), defined man as a composite of soul (mind) and body. And since he was desirous of "saving his soul" he had to "mortify" the flesh. Thus in deadening his body he progressively lost his life as he proceeded to gain the realm of soul and the whole world of ascetical spiritualities, including the pure vision of the metaphysical deity, that vanishing point of the world's navel. He saved his soul, i.e., his religious ego, at the price of his bodily life and of eros, for he wanted to be like God, a "pure spirit." But the ascetical idealist was not always successful in his impossible goal of disembodiment. Thus, when his body, ever yet alive, reasserted itself, the ascetic had to regard this as demonic. The ascetic, now turned sensualist, caricatures body as he caricatured soul. For neither soul nor body are experientially abstract, and thus in their state of abstraction spuriously pure or impure. In reality (and etymologically as well) the human psyche is neither a pure soul nor is it the object of scientific investigation. It is rather the life breath of the body; it is the body's own movement, particularly in the act of love. We cannot separate the life of the body from the body itself. The dualism is a specious one. Thus the flight from the body is a metaphysical adventures: the ascetic is simply "on a bad trip." The adventure is doomed from the outset; for, he is condemned to return, as an eroticist. It is the punishment for the pride of ego.

The body as the love body, as desired body (ἔρως), is not the object of erotic desire as wanton craving, as in the case of Don Juan, the libertine. The eros body is the human body as such, particularly as in living and loving encounter with the other, culminating in what we call the sexual experience. But this experience is not merely the experiencing of the beloved as an object of sensual contact but as the co-subject of an awakened awareness of what it means to be a human being, in Don Juan's case what it could mean to be a mature man, what ensuing fatherhood could mean. Physical love, as is well known, is no substitute for human love. But neither is the emaciated metaphysical love of soul for soul. Neither type of abstract love is satisfactory, be it love of soul for soul or of body for body. Both miss the human being. Both objectify and depersonalize the other. But of the two types of abstract love the metaphysical is probably the worst offender, because it passes as "higher" love than the physical. Physical love is regarded as "sinful"; but the greater sin is probably that of the spirit, for it is in bad faith in its encounter with the other, and thus any sort of crime against the other is possible. In such a state of bad faith one is able to say to those in need, "Be clothed and fed and depart in peace." But one does not clothe the needy nor feed them in reality. One feeds them with fine words instead. But words are cheap, especially with regard to love.

Sex is not just "physical"; neither is it "carnal." But it also need not be "spiritualized" or intellectualized. Rather, sexuality ought to be understood and accepted as simply human and "natural." The body is not first sexual and then bodily except for the ascetical mind. But the body, not as *a* body but as the embodiment of the human person, is first and foremost bodily and thus also sexual. This is emphasized in the recent vogue of "sensitivity training." Not every touch need immediately be interpreted in terms of eroticism, though it often is. People actually have to be trained to be able to accept the simple touch of another human being.[24] It is an indication of the alienation of members of our society that sensitivity training is necessary in times like ours. Modern man is still ascetically inclined: he withdraws from the touch of the other, or he is overly anxious about it. He needs to learn to overcome false eroticism or the fear of eros. The human body is not just an erotic physical object; it is the human person: somebody. We must begin, even in the realm of sexuality, to put sensual-eroticism in proper perspective. This will have the effect, if successful, of enhancing eros not of suppressing it. Eroticism is a form of eros suppression that must be overcome if we are to become "natural" human beings again.

Ontological sexuality means sex as it *is* without the overlay of eroticism wished on us by history. This history casts its long shadow even

---

[24] W. C. Schutz, *Joy, Expanding Human Awareness* (New York, 1967), *pass.*; cf. esp. ch. 2, "The Body", pp. 25 f.

over the twentieth century. Ontological sexuality is bodily. But as mentioned above, there is a split between male and female. This is an ontological difference, not just one championed by French poets, as in the well known phrase. The difference, moreover, is in each body, whether male or female, for the male has vestigial remnants of the female body and vice versa. The split is there but it is not absolute in a biological or psychological sense. Phenomenologically, one might say that the difference connotes a different shaping-up or "form" of the human being as such. A man is male in his major physical and psychological make-up and movement; and a woman also accordingly. But the minor movements and vestigial remnants can hardly be dismissed. And thus sexual problems become more involved. A man is not the "aggressor" in an absolute sense; neither is the woman "passive." What is naturally male and culturally masculine (to the extreme of a masculine body cult or mystique); what is naturally female and only culturally feminine (to the extent of a "feminine mystique")? The anxieties of the question become practical when the problem of homosexuality arises. Arthur Miller's *A View from the Bridge* is most enlightening on this subject. Suffice it to say that the truly heterosexual person need not be neurotically so. There is a pathetic fear of latent homosexuality among American males especially. Outright homosexuality arouses instant anxiety. Yet even the true heterosexual has a sufficient amount of the "female" in him, and this might even be psychologically important for him in a proper encounter with woman. Many supposedly tough, hard men may well be the embodiment of phallic mothers. The true heterosexual assimilates to his "male side" the masculinity of both father and mother and to his "female side" the femaleness of both. Thus he becomes convincingly masculine.

One might make some further distinctions here: between male and masculine on the one hand and female and feminine on the other. "Male" is the natural or "ontological" male, i.e., a man as he *is* without cultural and interpretative accretions from history, theory, etc. This does not, of course, mean man bereft of any culture at all, but freed from acculturative interpretations. It means man as a human being and thus as a bodily person, rather than as a physical entity to be studied and analyzed in the manner of Masters and Johnson (though that study is without a doubt significant from the scientific viewpoint). The same holds, mutatis mutandis, for woman, as female. "Masculine" and "feminine" (words borrowed from grammar?) are cultural labels for man and woman, as male and female. Thus go also their roles in a given society. These roles differ considerably in different cultures, though basically man and woman remain male and female as such. Hence in one culture the woman plows and the man sews; in another, like ours, the man plows and the woman sews. In Iran the man is emotional and the woman sober; in our culture the opposite obtains. Latins find no threat to their maleness in embracing one another openly. In our country only women may do this. Here we see a real distinction between the natural and the cultural. The "cultural" is not only the way things developed in a particular area; there are also negative elements, like unreasonable taboos. One of our taboos is the result of western asceticism ("Do not touch me"). An acceptance of all these factors and questions is important in counseling the subject who may be concerned about his or her sexual orientation, or indulges in projecting these fears into the movement of others. The sign of the true heterosexual is not just his ability to love woman in the encounter of eros but his capacity for true friendship, without fear, for members of his own sex. If his sexual movement is healthily heterosexual, i.e., other-directed (and the true other is a woman), he need not fear. The true heterosexual is not so compulsively; nor does he indulge in labeling others or deriding a way of life, which may be meaningful to those whose sexual movement seems to lie outside the scope of our own cultural categories.

The ontological body as dual, i.e., as male and female, is the ground of true sexuality. The body, as the human *somebody* or person, does not merely have sexual aspects but *is* sexual. In that the human person is split into the natural dualism of male and female, the urge toward reunion becomes a strong impulse. When writing of the ontological body philosophers usually mean the male body, as they have always spoken of the mind as that of a man. Thus it is important in this era of "women's liberation" to make sure that ontological body is understood in the sense of its natural dualism. Any philosophy has been "a man's game," originating in the homosexual circle of Plato's *Symposium* (cf. Alcibiades' complaint against Socrates!).[25] Woman was a domestic animal or she was a "companion" (ἑταίρα), a kind of ancient Greek geisha. In romanticizing woman Christianity really did no better, as regards allowing woman to be a human being. For from courtesan she was elevated to lady and was placed on a pedestal as a love object. And she was treated as such even by such lechers as Don Juan, who trans-

---

[25] E 210 f. Socrates seemed to have regarded the togetherness of διάνοια as more authentic than that of ἐρασταί. But this is not a question of the "spirit" vs. the "flesh."

lated love into sensual-erotic terms in order to be able to cope with her, as he copulated.

The ontological body is dual as male-female but not as soul-body. Woman is not the "soul" of the family and her husband its "body." The dualism of soul-body is traceable historically to the *Symposium* among other sources, and at least in its neo-Platonic form it is an attempt at transcendence within the male sublimation of ego. A cursory reading of the *Symposium* reveals that its topic is "Greek love." This is what makes neo-Platonic Christianity such a contradiction; for, though it is based on Platonic love, it gets indignant about Greek love. And yet Greek eros is the basis of that metaphysical love of which the Christian ro-manticists were so enamoured. As a substitute for the true dualism of male and female the transcendent other, the ego, was created within the male self, and projected outside (eventually becoming deified). But the psyche (as ego) is a poor substitute for the womanly other. The soul (*anima*, feminine gender) is the creation of the ascetical idealist who shuns encounter with the other, especially as woman, as this definite woman. Don Juan seems to have the same problem. He cannot find any lasting relationship with woman, hard as he tries or does not try, depending on your viewpoint. The problem of the ascetic is still a problem for our culture, it would seem. Whatever form it takes, that of classic religion, of scientific attitude, or of eroticism, as in the case of that inverted idealist, Don Juan, asceticism attempts to substitute a specious dualism (soul-body, mind-matter, etc.) for the real one: male-female. The split between mind and body begets a distorted sexuality and narcissism. The split between true opposites within the unity of the ontological body is the natural difference between man and woman, not the theoretical classic split between ego (mind) and self (body). The ontological body reflects both the difference and the unity of the sexes.

Most authors deal with sexuality predominantly from the masculine viewpoint, and in fact the author of this essay is writing on Don Juan not on Donna Anna or on Zerlina. As a result of this masculine emph-asis woman has been caricatured in literature, poetry, and in fact. She has been romanticized, exploited, debased. Thus the characteristic features of the female body have been caricatured to please the dis-torted imagination of the typical Don Juan. A mature approach to the human body, whether male or female, is seen in classic Greek sculp-ture, as well as in the romanticized works of Rodin. What impresses us about Greek sculpture is not sex but eros. Rodin, a modern, finds it useful to romanticize eros in that he presents the nude humans in the act of love. To the Greeks this would have been, so to speak, in poor taste, in that it was not classic form but too human. To Christians Rodin's work is either offensively or pleasurably erotic. Yet even in Rodin's sculptures, where figures do not pose heroically or in classic composure but are depicted in the actual movement of love making, the handsome bodies are not merely sensual-erotic. If they are "too human" rather than satisfactorily classic, they are none-the-less convincing. And they are beautiful.

Sensuality was introduced into civilization by Christianity. This post-biblical Christianity has been a historic mixture of neo-Platonism, medieval scholastic moralism, post-medieval cultures and subcultures, all of which could justly be designated by Nietzsche as "Platonism for the people." Sexual-eroticism, as portrayed in the person of Don Juan, is the natural outcome of the dichotomy of spirit and body. The life of the soul or of the mind is supposed to be man's higher life, the purer one, Descartes' *vrai moi*. The mind-matter dualism is a secularized form of the theological dichotomy of soul-body. As the ascetic feared letting his soul become contaminated by bodily contact with the other, so the philosopher makes sure that cognitive achievements are not touched by any emotive contents. Thus the scientist excludes all "subjectivity" from his objective researches. The spiritualist fears he will be dragged down by another's carnality. Yet if being in the world and dwelling on the earth means evolving our world and grounding it properly in eros, then the human being "is there" (Heidegger's *Da-sein*) also sexually, especially if he is there in the world with others (*Mitsein*). The anxiety of the classic ascetic is grounded in the apprehension that his ideological existence might be damaged or shattered by "physical contact." He fears he may have to dwell on the earth after all, whereas he has been conditioned to live a bodiless life or at least to aim at achieving it. Yet he is always pulled back, as it were by the gravity of the earth, with which the ontological body is one. Man, as the erect animal, is indeed transcendent in the sense that he stands up and thus can literally gain perspective in the world, unlike the animals. But his ignoble feet root him ever in the earth.[26] The ascetic is not satisfied to stand erect, he must take flight from the earth into a fantasy world, a better world, an otherworld in which the other plays no role.

Being in the world is not an abstract or static affair. One is *in* the

---

[26] R. Griffith, "Anthropodology: Man a-foot", in *The Philisophy of Body* (Chicago, 1970), ed. S. F. Spicker, pp. 273 f. (reprint from *Conditio Humana*, Berlin, 1966).

world in the sense that one evolves a world and dwells in it with others. This is understood best perhaps by lovers and poets. Ek-sistence means emergence, a standing-out, in the sense that man has stood erect, and thus is the "king of the animals." This is quite obviously phallic. The human male can be doubly erect, and his sexual erection (unless he is poor Schneider, as described by Merleau-Ponty) is directional as well as "intentional"; for it directs him toward coupling with the other and makes his whole being tend toward the opening up of the world of the other, thus toward the sharing of a bodily world. Intentionality is "lived" not in the sense that a metaphysical term is brought to life "in action," in this case in sexual interaction, but in the sense that our bodies tend toward one another, in that the male's body tends toward inwardness with his female partner. They thus are aware and alive together in a heightened sense, one possible only in union with one another. The man's body, i.e., the male as bodily and erect, is carried out and beyond himself toward the other. His sexual organ literally reaches out toward the other. In this sense man is "transcendent" in his being *as* bodily. He does not seek, like Don Juan, merely to possess a body; he seeks beyond himself the sharing of a mutual world. Certainly, physical intercourse is not of itself eros and does not automatically open up the world for us. Mere physical contact with its temporary excitement and fascination but its eventual frustration is the lot of Don Juan, even in marriage. If one does not let oneself go in the sense of letting go of neurotic control mechanisms, the act of love can never be fully satisfying. The "spiritualized" marriage is not the union of two in one flesh but the concurrence of two egos in sensual erotic contact. This concurrence is thus only physical intercourse; no bodily intersubjectivity results. Their union is soulless.

Don Juan is thus never satisfied either in soul or body. He must go on making love endlessly, because he can never make love adequately one single time. Marriage only complicates the problem for him. Don Juan, reared an idealist, regards the female body as a threat to his masculine ego; thus he dare not share a world with his partner. He must instead conquer her as an enemy. He must "have" her. Woman becomes a demonic power, both alluring and threatening at the same time. Even as he "sins" and enjoys the erotic pleasure as a reaffirmation of his ego, he feels the "evil" both of his own immoral act and of the presence of woman as the temptress, as the alleged corruptress of his idealism, as the seductress of his youth. Significantly his first corruptress, according to both Molina and Byron, was a married woman.

It is no small matter that his temptress was not a sordid witch or a siren but the frustrated wife of a Christian nobleman. The Don Juan problem is that of the ordinary warm-blooded Italian youth. Mozart understood this, and that is why his music (as *opera buffa*) and the Italian libretto of Lorenzo da Ponte are crucial for the understanding of Don Giovanni. Both Kierkegaard and Nietzsche described themselves as standing in admiration of Mozart's Italianate nature. And, of course, Einstein has informed us that he was indeed a lady's man.[27] When he was not in Italy Mozart always dreamed about it, Nietzsche relates. Don Juan is indeed Don Giovanni, the Latin Lover. Kierkegaard and Nietzsche could identify with Mozart; apparently Mozart could identify with Don Juan to some degree.

We see the Don Juan syndrome in any young man, who struggles with the false anxieties instilled in him by his ascetico-idealistic upbringing. Perhaps it is easy to see why idealism of this sort took root in such countries as Italy, where the *donne* are so beautiful, that they have to be more modest than elsewhere, and where the young men are so bodily it frightens them. The sun burns brightly in Italy, the land of *vino generoso*, and of the sensuous music of Vivaldi. Perhaps asceticism is a reaction against the overwhelming nature of life and of joy. There was fear one would be carried beyond return. Whatever the case, Don Giovanni, as the incarnation of anxiety and of music, is Italian. The syndrome becomes acute when such asceticism is exported to Germanic peoples. The Teutonic Don Juan, such as Nietzsche, becomes Faustian. It led to Nietzsche's severe critique of Christianity as such and to his attempt to reconcile good and evil, the eagle and the serpent. As Don Juan is the unacknowledged child of the ascetical monk, Nietzsche is the still unacknowledged prophet of Christianity gone astray.

The ascetic regards eros as morally threatening. But such morality is based on the unwarranted premise that the body is inferior to the soul and that it finds its place only in subservience to the spirit. The ontological body is thus demonized; and so the vicious circle of Platonic virtue spins on and on. Dualistic morality is doomed to this kind of short-circuiting. The only escape is a genuine return to that bodily unity we already are. Eventually suppressed body will emerge in demonic form; examples of this abound in literature and the arts. The sudden apparition of Satan and the colorful temptations of the classic recluse are

[27] A. Einstein, *Mozart, His Character, His Work* (Oxford, 1945), tr. Mendel, Broder, cf. ch. 3, "Mozart and the Eternal Feminine," p. 61f.

well known. The demon emerges as "il No al mondo" (Boito), i.e., as the negation of the angelic world, whose chorus sings too sweetly for him to be convinced. The recluse's fantasies are most revealing. The famous "Temptation of St. Anthony" which inspired the painter Grünewald, who in turn inspired the composer Hindemith, is replete with erotic fantasies, including anality, the sadistic, the macabre – all in bodily form. Body is animalistic.

But independently of Hieronymus Bosch or of Matthias Grünewald, or for that matter of Melle Oldeboerrigter,[28] Christianity has provided sufficient phallic symbols. The steeple of a church is an almost obscene phallus as it strives upwards for union with the heavens, symbolizing the wedding of earth and heaven in overt pagan manner. The lit candle, especially in a wreath, hardly needs commentary. When the candle is dipped in the font and is called on to impregnate the waters, as in the Easter ceremonies, this is more than a Christian festival. Christ emerging from the tomb, the baptized emerging from the waters, the purpled hierarch standing before his flock – all are phallic. Hidden phallic symbols abound. The ascetic would have to beware even of the cross. And surely he must not read the scriptures, where we are told that "God emptied himself. . ."

How did "God" figure in? The classic Christian deity had been the great Ascetic, the Pure Spirit who persuaded man to say No to the earth and to things bodily, in order to become godlike. This entailed remoteness from sexuality, and thus the greatest fear of the recluse was the "temptation of the flesh." One had to forego sharing a world with the other, if one wanted to draw nigh to this solipsistic deity, this false Lover, whose existence fed on the evacuated lives of his devotees. Yet the biblical God was not ascetical. He emptied himself in "kenotic" orgasm and begot a Son.* (The ecstasy of this God was not metaphysical but orgastic. The pagan gods were orgiastic, at least to the Christian mind.) In ecstasy the Son issued from the Father and was emptied forth into the world. This kind of mythology is to be found also in the early Greeks. For, the gods were said to "rain" as warm semen impregnating the earth and fructifying it, causing things to grow and abound.[29] Classic Christianity lost or discarded this earthy

---

* σαὐτὸν ἐκένωσεν

[28] "Melle's Melées" in *Avant Garde*, January 1969, p. 12 f.

[29] G. S. Kirk and J. F. Raven, *The Presocratic Philosophers* (Cambridge, 1960), p. 29. Here a passage from Aeschylus demonstrates the desire (ἐρᾷ ἔρως) between sky and earth; and the rain (ὄμβρος) from the sky fertilizes the earth and causes her to bring forth fruit. Sky and earth are described as bed partners (ἀπ' εὐνατῆρος οὐρανοῦ).

imagery. Current Christianity seems to becoming aware of it again

The ascetic aspired to a "higher love," and seemed not to know the "lesser." How one could love God if one had not loved man (in fact not just in idea) never seemed to become a practical problem for him. How one could say, one had loved man without ever having fallen in love, was a question that never arose. Falling in love would have been a fall from grace for the ascetic. But this fall could also have been interpreted as a fall from solipsistic posture, a free fall through the air and into the depths of love. Falling in love, not so much with love itself, but with a human being, is a test of fact rather than a metaphysical adventure. The ascetic was condemned to recoup a lost stage of bypassed love, and in the return to more humble beginnings he had to discover not what love was by definition, but whom he could love in encounter. The real tragedy of life is that human beings, made for love, never find it. Some historic shadow lies across their lives. Thus the sad spectacle of the frustrated recluse or the even sadder story of the frustrated spouses. It is a terrible thing to be made for love and never to have made love at all; but it is worse to have made love, like Don Juan, and never to have known the love of another and the sharing of a world.

"Post coitum omne animal triste." This was the strange judgment of the medieval ascetic. It was supposed to mean that after sexual union every animal, particularly the human, felt a sense of sadness and dejection, a sense of fall from grace and distance from deity. The ascetic felt superior to the common man. The animal, particularly the human one, could participate in sexual congress only under the condition that he feel guilt and sorrow, humiliation and anxiety. Indeed, Don Juan, the offspring of the ascetic, felt all this. How different eros! Instead of guilt, joy. Instead of humiliation and depletion, reunion with the earth and a sense of fulfillment. Sorrow is only for those who reject love or cannot find it. After intercourse the human partner feels not dejection and sadness but elation, wonder, peace, calm, tenderness, a sense of well-being that permeates his whole frame, renewed trust and strength. What kind of a deity would create a man to make love and then indoctrinate him with a faith that keeps him, whether as ascetic or as Don Juan, from ever finding love? Perhaps the deity needed to be rescued from the No-sayers, who in their resentment against life and love fashioned him in their own image. Some of these No-sayers became Christian models. Even such truly great human figures as Pascal and Francis of Assisi felt the blight of the No to life. The demonization of the human body and thus of human sexuality follows naturally from the introduction of sensuality into our culture. Sensuous music and sensual flesh militate against the spirit in a misconception of historic proportions. The spirit is conceived as the soaring eagle, and sensuality is an evil serpent. The serpent is, of course, another phallic symbol. The goal of the ascetic is to crush the head of the serpent either with the foot of the Virgin or in the talons and cruel beak of the eagle. In contrast to this Nietzsche's Zarathustra sought the reconciliation of the eagle and the serpent, his two favorite creatures. Thus the "Immoralist," who in reality sought to discover a more truly human morality. It was he who called mind "little reason" and the body "great reason."

Asceticism has entailed control of or negation of the procreative urge. There is a natural "asceticism" in marriage, which, of course, has little to do with any theological theories. It is rather an *askesis*, a "discipline" that teaches married partners to wait for one another, to bank the fires, to care for one another. For the wife is by no means just a vehicle of a man's passion; she is not the "remedium concupiscentiae" of the classic tradition. Rather, she is a human being, and thus she may not always be ready for sexual congress. It means that for some reason – not just the final weeks of pregnancy or some illness – a person must deny himself the sexual sharing of their communal world. In other words, he must say No, or No is said to him. But this negation is dialectical; there is always a return, an eventual Yes. In the dialectic of Yes and No the harmony of married life is to be found. The mistake of both the ascetic and of Don Juan is that the No of the former becomes a rigid way of life, and the Yes of the latter degenerates into selfish indulgence at the expense of the other. The ascetic said No and never returned to the source of life. He remained distant and aloof from his own sexuality and that of the other; thus he became spiritually desiccated and was lost in a sea of despair, all the while following a euphoric ideology of the spirit. The practical asceticism of marriage knows of the No but also of the joyous Yes. Antaeus always falls back upon the Earth, for Gaia restores him to life. He cannot struggle forever without this restoration to world and life. During his battle with the outside world Antaeus' energies are depleted. In falling back into the arms of Gaia he finds his manhood renewed. His union with Gaia is the warranty of his continued existence. The classic ascetic thought he could fight the battles of the soul without need of restoring himself on the bosom of the Good Earth. He did not realize that he was not saving but losing his "soul," i.e., his life, by saying No to the Earth, the

Source of Life. For he ended up half a man. Antaeus kept his manhood whole. But even in marriage it is necessary to remain "single." One does not simply fuse with the other. One remains oneself, even as one surrenders ego to the sharing of a world with the other. In order to preserve one's basic identity, the one must withstand the other, even as he stands with her.

One must remain "single" in marriage; but one can be single in the ordinary sense as well, i.e., unmarried, and this for any number of reasons. Brahms remained single for the sake of his music, though apparently, like Bruckner, he would have preferred to marry, had the situation been right. Mozart, of course, married, and had six children, two of whom survived. Christ apparently remained single for the "kingdom." But remaining single for whatever reason need not entail a rigid antisexual posture, though neither does it imply promiscuity. The single man may be aware of world like the married man, though it is difficult to see how he can be aware of it in any heightened manner. The classic mystics felt that their metaphysical ecstasies replaced genital encounter. Even the terminology of marriage was employed to convey this. But it is difficult to see the difference between genuine rapture and fantasy substitutes. Perhaps this can never be analyzed. Let those who want to marry do so; and let those who wish to remain single do so. There can be no absolute evaluation as to which is the "better" state; but as far as common experience can show, "Married is better," despite the attrition in the ranks of the espoused. Generally, the single state is precarious.

The phrase, "Married is better," is an improvement over the classic statement, "It is better to marry than to 'burn.' " As to "burning," eros is indeed a kind of "fire," which the Greeks thought to be divine, but Christians feared as detrimental to ascetical virtue. Eros was that vital energy of growth that made the difference between a boy and a mature man. It was that surging-forth (*surgissement*) that engaged the whole man in preparation for bodily encounter with the beloved and entry into the world. It was the mighty tide that carried a man forward into the sharing of a world and the procreating of new life therein. That this should have been distorted to mean mere "passion," when actually it contributed to the flowering of the whole person, is a problem in any attempt to evaluate classic Christianity in positive manner. By comparison with eros ascetical virtue is pale indeed; it is aloof, nervous, self-conscious. Eros can bring its fragile world tumbling down in a flash, as it attempts to make a human out of this being.

Virtue is solitary; eros seeks out the beloved. Men do not marry because otherwise they would burn with mere passion; they marry because eros brings them close to a kind of eternal fire which two can kindle and out of which life issues, as love is fostered. Apparently the "Father" burned with divine passion, so that he finally emptied himself and dying begot a "Son," a "Word made Flesh," and as fleshly meant to dwell among men. The Christian Logos, as neo-Platonic, was a spiritualization of the historic figure of Christ. A whole world separates this Logos from the *logos* of the Greeks as well as from the historic Revolutionary himself, who counted even prostitutes among his friends. The Christological era was a metaphysical adventure.

VI. CONCLUSION: DON JUAN AS VICTIM OF HISTORY

Don Juan is actually an engaging figure. His is not so much a historic tragedy as a tragic comedy. He is not so much the grand sexual conqueror, as he sees himself; rather, he seems to be the victim of his own exploits, which in turn are the result of his having been reared in the atmosphere of a historic dualism: idealism and sensualism. A crucial part of Don Juan's complex is not just his exaggerated fixation on woman in the modality of sensual eroticism but also his distorted view of man. The other side of his overt and frenetic heterosexuality is his latent homosexuality. Kierkegaard maintains that all the characters of Mozart's opera have an erotic relationship with Don Juan, and that especially Leporello, his servant, bears witness to this bond in the music itself. He is bound to Don Juan almost against his will, Kierkegaard writes, and on this very account one hears Don Juan himself singing in and through the person of his man servant.[30] We need not characterize this relationship as latently homosexual, this especially if there is such a thing as true friendship and service. But Don Juan is incapable of friendship; he is not free to be anyone's friend. He can only have an erotic relationship with people, and he exploits them as such. Don Juan himself is not a free man; he is in bondage to sensual eroticism; and thus he can experience neither eros nor friendship. And since music as such is also experienced by Christian culture as sensuous, all this is brought out in the musical drama of Mozart, the "crown of all opera" in the judgment of Kierkegaard.

Don Juan's sexual anxiety expresses itself consciously as flight from woman but in the modality of overt pursuit of female eroticism. This is

[30] *NMA*, II, 5, 17, p. 284 f.

a phenomenon observable in a good many average males, who are unsure of their maleness and thus must assert their masculinity. We see here two sides of sensual-eroticism: false heterosexuality and latent homosexuality. But the latter is hardly unexpected, when one considers that the Platonic Academy itself was quite inverted, and that the flight to the otherworld which it launched and which Christian ascetics capitalized on was part of this false movement. The spiritual muscle-man, the mental gymnast, is not altogether unlike his physical counterpart. Neither is a model of humanness or of classic proportion and balance. Both represent a distortion and perhaps what Hölderlin called "an untimely growth." Yet this is the story of our civilization, insofar as it has been affected by ascetical idealism and metaphysics. The spiritual and the intellectual supermen are set up as models, whereas in reality they are monstrous and distorted, as lacking in beauty and balance as the modern phenomenon of the muscle-man or the beach boy.

Mere physical sexuality is a caricature; it is a derived and very deficient form of the ontological body as love body. The deficiency consists in making the subject an object of lust. Our human body underlies (sub-ject) all we are and do; it is not something over against us (ob-ject) that we observe in a scientific study or which we stare at like a voyeur. The "physical" body is a mental trap. Don Juan, the idealist, is fascinated with and fixated upon his own physical prowess. His outgoing movement never really transcends this stage. And thus even in his frantic pursuit of woman he remains captivated and victimized by his own sexual egotism. A true *askesis* needs to be worked out that would provide the basis for therapy,[31] so that the Don Juan syndrome might give way to an understanding of the eros body. Traditional asceticism bracketed the body and caused it to be regarded as physical. Don Juan is the victim of this historic theory. A true asceticism does not go beyond eros body but discovers it and abets the growth of love between sexual partners. Don Juan needs to transcend both classic and ascetical idealism and sensual eroticism. He needs to discover the meaning of an existential eros in the true encounter with the other. And this encounter takes place not in the shadow of anxiety but in the joy and wonder of loving surrender.

[31] Both Greek words provide interesting historical meaning-models. Ἄσκησις means "exercise" and practice in the acquiring of a skill, in this case the art of love. It would entail "asceticism" in the sense that this would imply ridding ourselves of any historic overlay-meanings, such as eroticism, but also any theologisms or romanticism based on historic metaphysical idealism. The word θεραπεία has to do not with "therapy" but with growth. Cf. Plato's *Theages*, B 6, where it designates the growth of plants and also refers to man as such. Eros *therapy* would imply learning to grow in love.